'The focus of the book on bioeconomy is highly appropriate at a time when we are faced with the global grand challenges and an ambitious target to meet the sustainable development goals. The book provides an excellent review of the bioeconomy concept, its potentials and constraints, drawing experiences from several countries, covering various disciplines and sectors including agriculture, forestry and fisheries.'

Olaug V. Bollestad, Minister of Agriculture and Food, Norway

'This timely book provides potential pathways for successful implementation of sustainable bioeconomy, through comprehensive coverage of relevant sectors in various parts of the world. This is in line with United Nations University's expectations on bioeconomy to be a resort for environmental, social and economic sustainability.'

Taikan Oki, Senior Vice-Rector, United Nations University, Japan

'This book makes a timely and innovative contribution towards understanding the relevance and practicality of bioeconomy, its concept, principles and best practices, towards achieving Sustainable Development Goals, including Zero Hunger. It deserves to be read and used widely!'

Hans Dreyer, Director, AGP, Food and Agriculture Organization of the United Nations

The Bioeconomy Approach

This book examines the bioeconomy concept, analysing the opportunities it can generate, the constraints and the potential benefits for society.

The main objective of bioeconomy is to promote economic development, by creating jobs and enhancing the sustainable utilization of bio-resources. A primary driver of bioeconomy strategy, therefore, is the need to respond to the growing population's food and economic requirements. While today research and literature related to bioeconomy are limited, this book presents a unique collection of perspectives on the complex dimensions of the bioeconomy debate. Drawing on the experiences from Europe, Asia and Africa, it presents an international overview. The chapters address a wide range of issues, including coastal-land interactions, ecosystem services, food production, rural development, agriculture, forest management and bioenergy. As a whole, the volume outlines what role bioeconomy can play in contributing to the United Nations Sustainable Development Goals (SDGs) without compromising on the ecological sustainability and equitable distribution of benefits. The book concludes by providing recommendations for developing bioeconomy in respective sectors (agriculture, forestry, fisheries, renewable energy) and directions for planning future bioeconomy programmes and strategies.

The Bioeconomy Approach will be of great interest to students and scholars of ecological economics, development economics and environmental economics, as well as policy-makers and practitioners involved in sustainable development.

Udaya Sekhar Nagothu is Research Professor and Director of the Centre for International Development (CID) at the Norwegian Institute of Bioeconomy Research, Ås, Norway.

Routledge Studies in Food, Society and the Environment

Localizing Global Food
Short Food Supply Chains as Responses to Agri-Food System Challenges
Edited by Sophia Skordili and Agni Kalfagianni

Seafood Supply Chains
Governance, Power and Regulation
Miriam Greenwood

Civil Society and Social Movements in Food System Governance
Edited by Peter Andrée, Jill K. Clark, Charles Z. Levkoe and Kristen Lowitt

Voice and Participation in Global Food Politics
Alana Mann

Plant-Based Diets for Succulence and Sustainability
Edited by Kathleen May Kevany

Sustainable Food System Assessment
Lessons from Global Practice
Edited by Alison Blay-Palmer, Damien Conaré, Ken Meter, Amanda Di Battista and Carla Johnston

Raw Veganism
The Philosophy of the Human Diet
Carlo Alvaro

The Bioeconomy Approach
Constraints and Opportunities for Sustainable Development
Edited by Udaya Sekhar Nagothu

For more information about this series, please visit: www.routledge.com/books/series/RSFSE/

The Bioeconomy Approach

Constraints and Opportunities for Sustainable Development

Edited by Udaya Sekhar Nagothu

Routledge
Taylor & Francis Group
LONDON AND NEW YORK

earthscan
from Routledge

First published 2020 by Routledge

2 Park Square, Milton Park, Abingdon, Oxon OX14 4RN
605 Third Avenue, New York, NY 10017

Routledge is an imprint of the Taylor & Francis Group, an informa business

First issued in paperback 2021

British Library Cataloguing-in-Publication Data
A catalogue record for this book is available from the British Library

Library of Congress Cataloging-in-Publication Data
Names: Nagothu, Udaya Sekhar, editor.
Title: The bioeconomy approach : constraints and opportunities for sustainable development / edited by Udaya Sekhar Nagothu.
Description: New York, NY : Routledge, 2020. | Includes bibliographical references and index.
Identifiers: LCCN 2019054108 (print) | LCCN 2019054109 (ebook) | ISBN 9780367335717 (hardback) | ISBN 9780429320651 (ebook)
Subjects: LCSH: Sustainable development. | Sustainable agriculture–Economic aspects.
Classification: LCC HC79.E5 .B5652 2020 (print) | LCC HC79.E5 (ebook) | DDC 338.9/27–dc23
LC record available at https://lccn.loc.gov/2019054108
LC ebook record available at https://lccn.loc.gov/2019054109

ISBN: 978-0-367-33571-7 (hbk)
ISBN: 978-1-03-217330-6 (pbk)
DOI: 10.4324/9780429320651

Typeset in Sabon
by Wearset Ltd, Boldon, Tyne and Wear

Contents

Figures

Tables

Boxes

Contributors

Alma Linda M. Abubakar is the Programme Development Officer of the FAO Regional Integrated Pest Management/Pesticide Risk Reduction Programme, at the FAO's Regional Office for Asia and Pacific in Bangkok, Thailand.

Masahiro Aiba is Assistant Professor at the Research Institute for Humanity and Nature, Kyoto, Japan.

Erhan Akça is Professor at the School of Technical Sciences, Adiyaman University, Turkey.

Yuka Bando is a Masters student at the Graduate School of Agricultural Science, Kobe University, Japan.

Suha Berberoğlu is Professor, in the Department of Landscape Architecture, the University of Çukurova, Adana, Turkey.

Giacomo Branca is Associate Professor, in the Department of Economics, Engineering, Society and Business, Tuscia University, Italy.

Luca Cacchiarelli is a Researcher, in the Department of Economics, Engineering, Society and Business, Tuscia University, Italy.

Vornthalom Chanthavong is a Farmers Field School specialist (Strengthening Agro-Climatic Monitoring and Information System (SAMIS)), Vientiane, Lao PDR.

Hugo Despretz is an agronomist at VegInnov, specializing in protected cultivation of vegetables in the tropics, France.

Ngo Tien Dung is Senior Advisor at the Centre of Initiatives on Community Empowerment and Rural Development (ICERD), Hanoi, Vietnam.

Selim Kapur is Professor (Emeritus), in the Department of Soil Science and Plant Nutrition, University of Çukurova, Adana, Turkey.

Jan Willem Ketelaar is Chief Technical Advisor of the FAO Regional Integrated Pest Management/Pesticide Risk Reduction Programme, at the FAO's Regional Office for Asia and Pacific in Bangkok, Thailand.

Kunihiko Kobayashi is Researcher at the Research Institute for Humanity and Nature, Kyoto, Japan.

Hiroko Kurokawa is Senior Researcher at the Forestry and Forest Products Research Institute, Tsukuba, Japan.

Roar Linjordet is Senior Researcher, in the Division for Environment and Natural Resources, the Norwegian Institute of Bioeconomy Research, Ås, Norway.

Primrose Magama is Junior Research Engineer, in the Agricultural Engineering, Agricultural Research Council, at the Institute for Agricultural Engineering, Pretoria, South Africa.

Daisuke Maki is President of A0 Inc., Kageishi, Nishiawakura, Aida, Okayama, Japan.

Shunsuke Managi is Professor at the Urban Institute, Kyushu University, Fukuoka, Japan.

Armel Clement Goudjo Mensah is the Head of the Vegetable Programme at the National Institute of Agricultural Research of Benin (INRAB), Benin.

Abha Mishra is the Director of the ACISAI, Asian Centre of Innovation for Sustainable Agriculture Intensification, Klong Luang, Thailand.

Salome Modiselle is an agricultural economist at the Agricultural Research Council , the Economic Analysis Unit, Pretoria, South Africa.

Mokhele Moeletsi is Research Team Manager, Agrometrology, at the Agricultural Research Council, Soil, Climate, Water, Pretoria, South Africa.

Takanori Nagano is Associate Professor at the Graduate School of Agricultural Science, Kobe University, Japan.

Udaya Sekhar Nagothu is Research Professor and Director (Centre for International Development) at the Norwegian Institute of Bioeconomy Research, Ås, Norway

Tohru Nakashizuka is Professor at the Research Institute for Humanity and Nature, Kyoto, Japan.

Thibault Nordey is a specialist in tropical vegetable cropping systems modelling, at the World Vegetable Centre, Eastern, and Southern Africa, Arusha, Tanzania, and CIRAD – UR HortSys, Montpellier University, Montpellier, France.

Michio Oguro is Senior Researcher at the Forestry and Forest Products Research Institute, Tsukuba, Japan.

Yutaka Okano is the Director of A0 Inc., Kageishi, Nishiawakura, Aida, Okayama, Japan.

Takeo Onishi is Associate Professor at the Faculty of Applied Biological Sciences, Gifu University, Japan.

Chiara Perelli is Junior Researcher in the Department of Economics, Engineering, Society and Business, Tuscia University, Italy

Avakat Phasouysaingam is National Investigator at the Faculty of Agriculture, National University of Laos.

Anne Strøm Prestvik is a Researcher at the Norwegian Institute of Bioeconomy Research, Ås, Norway.

Trond Rafoss is Associate Professor in the Department of Natural Sciences, at the University of Adger, Norway.

Divina Gracia P. Rodriguez is a Researcher, in the Division of Food Production and Society, at the Norwegian Institute of Bioeconomy Research, Ås, Norway.

Mayling Flores Rojas is Mechanization Officer, at the Food and Agriculture Organization of the United Nations, Rome, Italy.

Takahiro Sasai is Assistant Professor in the Department of Geophysics, Graduate School of Science, Tohoku University, Japan.

Rei Shibata is Assistant Professor at Niigata University, Japan.

Takayuki Shiraiwa is Associate Professor at the Institute of Low Temperature Science, Hokkaido University, Japan.

Peter Sprang is Technical Coordinator, Sustainable Rice Platform (SRP), at the International Rice Research Institute, Los Banos, the Philippines.

Mehreteab Tesfai is Senior Researcher at the Norwegian Institute of Bioeconomy Research, Ås, Norway.

Takahiro Ueyama is Village Counsellor, Kageishi, Nishiawakura, Aida, Okayama, Japan.

Preface

At a recent UN summit, one of the world leaders asserted: 'For us, sustainability is the navigation instrument, the compass into the future.' Unless sustainability becomes a fundamental criterion in policy and practice in the future, protecting the Earth's resources for future generations will not be possible. We have no other option but to follow the sustainable development path to address the grand global challenges we are facing today. At the same time, we must be optimistic, as it is necessary to sustain and improve the quality of life, mitigate future climate risks, and protect the environment. The sustainable bioeconomy approach has gained importance over the last decade among scientists and policy-makers as one of the most promising alternative paradigms to address the sustainable development goals. However, one must be careful with the environmental and social risks associated with new paradigms. The fundamental premise should be to ensure healthy livelihoods in a healthy ecosystem.

This interdisciplinary book attempts to clarify the bioeconomy concept and analyses the associated risks and opportunities. It further highlights the multi-dimensional benefits of sustainable bioeconomy development, including circular bioeconomy. As the impacts of the bioeconomy spread beyond local and regional borders, a common agenda is necessary to keep a balance between the economic, environmental and social sustainability goals.

The various chapters in the book drafted by 41 experienced researchers and consultants from several disciplines, representing 28 agencies worldwide, bring together diverse experiences. The book covers cases from biomass production from the ocean, agriculture, forestry and bioenergy. Several chapters in the book highlight relevant sustainability indicators, including the Human Development Index, net primary production, the local ecological footprint, soil organic carbon and other ecosystem services that can help in monitoring sustainability impacts of bioeconomy initiatives. The book will be useful to a wide range of audience including scientific community, development agencies and policy-makers.

Acknowledgements

I would like to thank all the contributors of the various chapters in the book, particularly the lead authors who have spent time coordinating and drafting the chapters.

This book would not have been possible without support from the Norwegian Institute of Bioeconomy Research, Ås, Norway, and Kobe University, Japan, that hosted me during my short sabbatical period in the autumn of 2019.

I would also like to thank Dr Takanori Nagano for his support during the compilation of the book manuscript.

Abbreviations

ACISAI	Asian Centre of Innovation for Sustainable Agriculture Intensification
AD	anaerobic digestion
AIT	Asian Institute of Technology
APHRODITE	Asian Precipitation–Highly-Resolved Observational Data Integration Towards Evaluation of Water Resources
ASC	Aquaculture Stewardship Council
AWD	Alternate Wetting and Drying
BBE	bio-based economy
BCDP	the Bioeconomy Community Development Programme
BCR	benefit-cost ratio
BIC	Bio-Based Industries Consortium
BID	bioeconomy domain
BTP	the Bioeconomy Transformation Programme
CASA	Carnegie–Ames–Stanford Approach
CAZ	carbon accumulation zones
Cb	carbon
CE	circular economy
CID	Centre for International Development
CSA	climate-smart agriculture
CSO	civil society organization
DF	discount factor DF
DHA	docosahexaenoic acid
DOM	dissolved organic matter
DST	Department of Science and Technology
EF	Ecological Footprint
EPA	eicosapentaenoic acid and
ETG	Export Trading Group
FAO	Food and Agriculture Organization
Fe	iron
FFS	Farmers Field School
FISP	Farm Input Subsidy Programme
FIT	Feed-in Tariff

FOLU	Food and Land Use Coalition
GA	Göksu anthroscape
GCF	Green Climate Fund
GDP	gross domestic product
GEF	Global Environment Fund
GFBF	Giant Fish-Breeding Forest
Gha	global hectares
GHG	greenhouse gases
GI	Geographic Indication
GTZ	German Agency for Technical Cooperation
HA	humic acid
HDI	Human Development Index
HNLC	High Nutrient Low Chlorophyll
ICERD	Centre of Initiatives on Community Empowerment and Rural Development
IFES	Integrated Food and Energy Systems
IKS	indigenous knowledge systems
INRAB	Institute of Agricultural Research of Benin
IPBES	Intergovernmental Science-Policy Platform on Biodiversity and Ecosystem Services
IPM	Integrated Pest Management
IRR	Internal Rate of Return
IRRI	International Rice Research Institute
IUCN	International Union for the Conservation of Nature
IWG	Inter-Ministerial Working Group on Bioeconomy
KBBE	Knowledge-Based Bio-Economy
LE	linear economy
LEDs	light-emitting diodes
LLDPE	linear low-density polyethylene
LMB	Lower Mekong River Basin
MoAIW	Ministry of Agriculture, Irrigation and Water development
N	nitrogen
NASFAM	Commercial National Smallholder Farmers' Association of Malawi
NDVI	normalized difference vegetation index
NGO	non-governmental organization
NPIW	North Pacific Intermediate Water
NPK	phosphorus and potassium mix
NPP	net primary productivity
NPV	net present value
P	phosphorus
PC	protected cultivation
PCA	principal component analysis
PCV	protected cultivation of vegetables
PES	payments for ecosystem services

POPs	persistent organic pollutants
PP	payback period
PPP	public–private-partnership
RIHN	Research Institute for Humanity and Nature
RRI	Regional Rice Initiative
RRS	Regional Rice Strategy
S&G	Save and Grow
SAMIS	Strengthening Agro-Climatic Monitoring and Information System
Si	silicon
SI	Sustainable intensification
SIRP	sustainable intensification of rice production
SME	small and medium-sized enterprise
SNV	the Netherlands Development Organization
SOC	soil organic carbon
SRP	Sustainable Rice Platform
SSA	Sub-Saharan Africa
TOC	total organic carbon
UV	ultra-violet
VC	value chain
VSLG	village saving and loans group
WFD	Water Framework Directive

1 The bioeconomy approach and sustainable development

A review of the concept, opportunities and constraints

Udaya Sekhar Nagothu and Takanori Nagano

Introduction

Global challenges, including climate change, food insecurity, economic crises, lack of jobs and political conflicts are driving people to migrate to cities. Outmigration is already changing the social landscape, the population dynamics and politics in various countries in both the developed and developing world. In the long term, this will lead to distress in rural areas and over-population and unemployment in the urban and peri-urban areas. Urbanization may lead to energy savings, efficient infrastructural investment, increased communication and innovation. Thus, the challenges that stem from the rapidly changing physical and social landscapes must be carefully considered in future planning. To address the growing problems, equitable social and economic development and the creation of job opportunities with minimum impact on the environment should be a priority for planners and government agencies (Beddington et al., 2012).

Population growth and the current patterns of production and consumption are fast exhausting the natural resources, degrading ecosystems, and generating waste and pollution at an unprecedented rate. According to some estimates, there will be a need to produce 50 per cent more food and energy to meet the growing demand by 2030 (FAO, 2016a). The mantra of business as usual will no longer be able to address the growing demands and associated global challenges. Future development must be based on strictly responsible, accountable, and sustainable use of natural resources. This brings 'sustainable development' back onto the policy agenda at both the international and national levels of most countries. The United Nations Sustainable Development Goals (SDGs), defined in 2015, thus, provide political legitimacy to national governments to integrate them into their national development plans (UN, 2015).

The United Nations (UN) Sustainable Development Goals

Since the 1950s, there has been significant technological and economic progress on many fronts, but the development has been highly uneven and unsustainable. Addressing the audience at the UN Summit in New York, where 193 countries agreed to achieve the SDGs, the German Chancellor Angela Merkel rightly said: 'For us, sustainability is the navigation instrument, the compass into the future.' The UN believes that achieving SDGs will bring about equitable development across regions and various social groups (UN, 2015). However, there are critics who do not agree that SDGs and the targets set by the UN are easily achievable and believe that they will not be able to provide any legitimate framework for cooperation (Easterly, 2015). However, others are positive that sustainable use of bioresources can help to achieve most of the SDGs that in turn can improve social, economic and ecological conditions (Dietz et al., 2018).

The SDGs in general provide economic and political legitimacy to introduce sustainable business initiatives, based on natural resources, especially in developing countries (Calestous, 2016). However, conflicting national priorities, particularly in developing countries with limited resources, make it hard to divert investments to promote new initiatives. Countries must see addressing SDGs as an opportunity, where new development paradigms such as bioeconomy can play an important role, and, in the process, support the global consensus and provide an international commitment to sustainably manage Earth's resources.

Bioeconomy: definition, scope and development

The European Commission (EC) defined 'bioeconomy' as 'the production of renewable biological resources and the conversion of these resources and waste streams into value added products, such as food, feed, bio-based products and bioenergy' (EC, 2012). The Food and Agricultural Organization (FAO) defined bioeconomy as 'the knowledge-based production and utilization of biological resources, biological processes and principles, to sustainably provide goods and services across all economic sectors' (FAO, 2016b). Another definition considers 'bioeconomy as the sustainable utilization of renewable resources for economic, environmental, social, and national security benefits' (Golden and Handfield, 2014). These definitions have a common terminology emphasizing the *sustainable use of biological or renewable natural resources*. According to Bell et al. (2017), sustainable bioeconomy has the potential to contribute to climate change mitigation, with oceans, forests and soils being major carbon sinks and fostering negative carbon emissions. This book, and particularly this introductory chapter, will attempt to discuss some of the intriguing questions that scientists and development agencies are facing today in the bioeconomy debate: *Is sustainable bioeconomy the right paradigm to address grand global challenges? Will it be able to generate jobs and economic growth without any negative impacts on the environment?*

According to the Organisation for Economic Cooperation and Development (OECD, 2009), bioeconomy is a knowledge-driven concept that has the potential to contribute to significant global economic, social and environmental challenges in an integrated framework. Bioeconomy is now the European Union's (EU) response to addressing the key global environmental challenges (European Commission, 2013). If managed in a sustainable manner, bioeconomy can simultaneously provide a wide range of public goods, including ecosystem services, as well as providing jobs and promoting new business opportunities. The concept, however, overlaps with similar approaches, such as the 'green economy', but also differs in the sense that it is more regulatory in its approach, and perhaps emphasizes the technical and economic sustainability more and social sustainability less. Recent studies have identified that bioeconomy is a key component in the global sustainability transition (El-Chichakli et al., 2016; Kircher, 2014). However, to achieve sustainable bioeconomy, technical, economic, and social prerequisites are necessary that the bioeconomy itself cannot create (Pfau et al., 2014). This is a huge challenge for developing countries with limited finances, lack of access to technology and inadequate social preparedness. Overall, bioeconomy development will depend on technology advances and sufficient biomass availability (Scarlat et al., 2015).

There are several factors or drivers, including population growth, climate change, food insecurity, increased demand for biofuels and trade, that have influenced governments to adopt the bioeconomy development path so far. The primary driver of bioeconomy in developing countries is the need to respond to their growing population's food and economic requirements, whereas the primary driver is to reduce greenhouse gas emissions (GHGs) and negative environmental impacts in developed countries, as well as to generate economic growth. A study by Langeveld et al. (2010) concluded that the over-use of non-renewable fossil-based resources, and the serious environmental impact this has caused over the years, have become one of the primary drivers for bioeconomy. According to Dietz et al. (2018), the reasons for adopting bioeconomy could be one or a combination of several that include: (1) reducing use of fossil fuels and replacing them with renewable bio-based resources; (2) increasing productivity in bio-based primary sectors (e.g. agriculture, fisheries, forestry, bioenergy); (3) increasing efficiency in biomass use; and (4) to support value chain enhancement of products that could lead to more job opportunities. One of the main challenges will be the replacement of fossil fuels with renewable energy sources that has been happening at a slow pace and is likely to influence the other factors. This can be attributed to lack of follow-up, lack of technology, poor investments and industrial or political lobbies opposing the reduction of fossil fuels.

An associated term, the 'circular economy' (CE) has been gaining importance recently. CE can be defined as 'the value of products, materials and resources maintained in the economy for as long as possible, thus reducing the dependence on new raw materials and the generation of waste minimized' (EC, 2015). Ideally, a CE model reflects the need for the world's population to considerably reduce its material footprint and consumption of natural resources, as described in the UN Sustainable Development Goals (UN, 2015). The two approaches, CE and bioeconomy, are different, but overlap in some respects. CE introduces a differentiation between durable and consumable parts of a product. While durable parts should be maintained in the economy, consumables in the CE should be biological ingredients or at least non-toxic ones, which can go back into the biosphere or be used in the cascade of consecutive uses (Ellen MacArthur Foundation, 2013). CE also demands renewable energy be used to decrease resource dependence (ibid.). The term 'circular bioeconomy' is now being used by the scientific community, business enterprises and policy-makers.

Circular bioeconomy

The 'circular bioeconomy' is defined as the intersection between bioeconomy and CE (Carus and Dammer, 2018). The assumption is that the circular bioeconomy will increase resource use efficiency, reduce waste along the value chain, promote nutrient cycling, and provide access to basic services and decent jobs for a better quality of life (Su et al., 2013; Kalmykova et al., 2018). According to Carus and Dammer (2018), the organic resources and waste from agriculture, forestry, fishery, food and feed sectors can only be integrated into the CE through bioeconomy, while the latter will hugely profit from increased circularity. In the process, integrating and engaging the value chain (VC) actors, from producers, manufacturers, intermediaries, traders and retailers to the final consumer in the production-consumption path, will be crucial. However, the costs may escalate as (bio)economies seek to expand. These involve not only 'circulation costs', in effect, recycling and reproduction costs but also vastly increased transaction costs. Technological advances may reduce both types of costs, but that may take time.

Figure 1.1 broadly illustrates the concept of bioeconomy. Bioresources originate in nature, which means bioeconomy is dependent on the health of various ecosystems. Therefore, sustainable production/harvest from ecosystems and the safe release of waste back into ecosystems are primary conditions for sustainable bioeconomy. Energy and materials are required in large quantities and, thus, they compose the basis of sustainable bioeconomy, mainly replacing fossil fuels. Food, feed and medicine have more value with less quantity. For more value addition, differentiation with knowledge and technology is required. In the sustainable bioeconomy

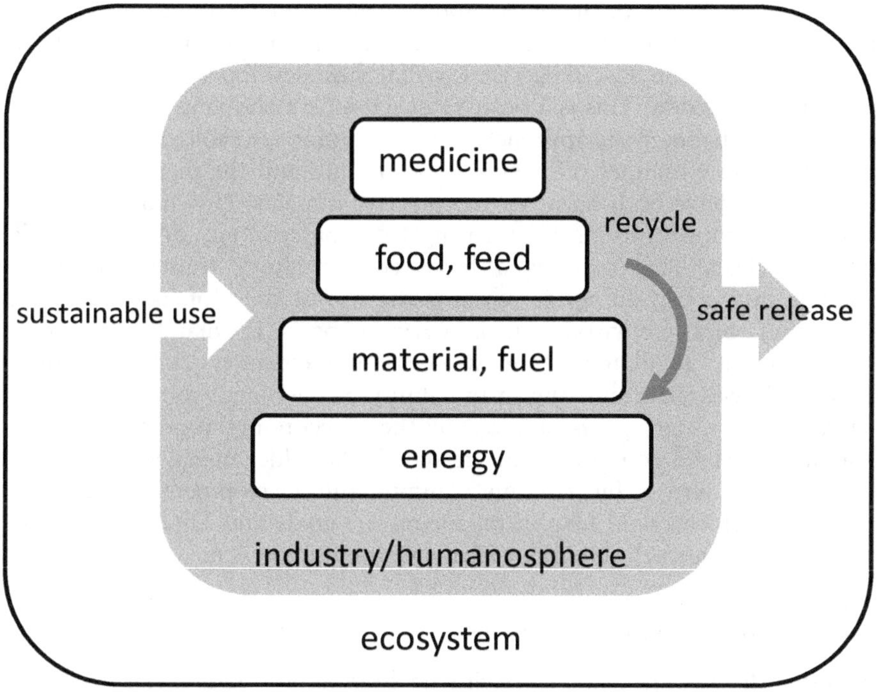

Figure 1.1 The concept of sustainable bioeconomy.

process, some of the waste stream must be recycled and the rest safely released back into the ecosystem.

The 'circular bioeconomy', on the other hand, is still a new concept, and how much it can be put in practice is debatable (Carus and Dammer, 2018). Therefore, the need to develop evidence-based approaches must be recognized to acknowledge the benefits of the circular bioeconomy. The EU has developed a new action plan based on the circular bioeconomy approach that will launch several pilot actions across the EU (EC, 2018). Following this, a few countries within the EU, starting with the Netherlands, have already set a target of becoming 50 per cent 'circular' by 2030 and 100 per cent by 2050 (*Independent*, 2019). This book will focus primarily on the sustainable bioeconomy approach and the use of bio-based resources and their potential to ensure sustainable development.

Global initiatives promoting the bioeconomy approach

According to a comprehensive study conducted by El-Chichakli et al. (2016), nearly 40 countries are trying to boost their bioeconomy in various sectors. The EU and the OECD have been at the forefront in providing the political momentum for the bioeconomy agenda and simultaneously taking initiatives to promote international cooperation (TransNational Institute, 2015). Outside the EU, it is the US that has developed a broad-based bioeconomy agenda. This is a positive development showing that countries are open to change their approach to use and manage biological resources in a sustainable manner. However, bioeconomy and the increased use of biomass will have both negative and positive impacts. How countries will address the constraints or risks arising from bioeconomy development is still not clear. This requires a balanced regulatory framework (both enabling and constraining) to monitor the use of the biomass and check the impacts on the environment on a regular basis. Ultimately, the extent to which the regulatory framework is implemented and monitored responsibly will determine the sustainability of the bioeconomy. A balance between the primary goals of adopting the bioeconomy approach is desirable and must be consciously addressed. Stakeholder engagement should be an integral part of bioeconomy planning and development, as observed in the EU, which could help in balancing the goals and implementing the regulatory framework to check the environmental impacts.

The EU and bioeconomy development

In 2009, the OECD came out with its bioeconomy strategy, but it was narrow in its focus with the emphasis on biotechnology (OECD, 2009). Then, in 2010, Germany became the first country to launch a 'National Research Strategy for Bioeconomy 2030' (Georg, 2018). The EU followed by launching its first Bioeconomy Strategy in 2012, covering various sectors such as energy, forestry, water, agriculture and marine resources (EC, 2012). The Strategy was supported by an action plan containing detailed measures with emphasis on research, innovation and skills, stakeholder engagement and enhancement of markets and competitiveness in bioeconomy. The EC further developed a 'circular bioeconomy plan' (EC, 2015). After a review in 2018, the EC updated the 2012 Bioeconomy Strategy, with an emphasis on scaling up the bio-based sectors, unlocking investments and markets, and deploying local bioeconomy initiatives across the EU, with due consideration to the ecological boundaries (EC, 2018).

The Knowledge-Based Bio-Economy (KBBE) was an offshoot of the EU's life sciences research agenda of the 1990s. KBBE aims to make agriculture and fisheries more sustainable and efficient. The European Commission's research priorities on bioresources are now based on the KBBE, a new political-economic strategy that plays a key role in shaping relevant policies and institutional arrangements with the aim of creating 'sustainable capital' (TransNational Institute, 2015).

Some of the EU member states, including the Netherlands, Belgium, Finland and others, have developed their own national bioeconomy strategies in line with the EU strategy. Country plans and their implementation are currently supported by funds from the EC in addition to their own national investments. Bioeconomy development, initiated in Germany since 2010, has shown significant success that was possible due to the politically coherent framework developed jointly by the relevant ministries, including Agriculture, Economic Affairs, Environment and Foreign Affairs, and supported by a 2.4 billion Euros investment (Georg, 2018).

The Horizon 2020 programme, one of the most elaborate global research and innovation programmes promoted by the EC, supports innovative research in bio-based economy (EC, 2013). The programme supported research on food security, sustainable agriculture and forestry, marine, maritime and inland water research with a focus on topics such as blue growth, rural development and bio-based materials (EPRS, 2017). The programme encouraged science-business linkage, public-private partnerships in different sectors and capacity-building within scientific institutions and business enterprises. Under this programme, several pilot actions were initiated in rural, coastal and mountainous areas for sustainable food and farming systems, forestry and bio-based products and these are proving to be successful. The successful initiatives are now being scaled up across the EU and beyond. Research and innovations from the EU initiatives can provide a learning curve to other countries who have initiated bioeconomy strategies.

The Nordic Bioeconomy Strategy

The Nordic Bioeconomy Strategy combines environmental, social and economic ambitions for a sustainable region (Nordic Cooperation, 2018). The strategy further emphasizes applying a cross-sectoral approach to optimize the use of resources. The Nordic Bioeconomy Strategy follows the EC approach in many ways, particularly the promotion of public-private partnerships involving industry, government agencies and research institutions that is crucial in creating a successful bioeconomy model (Nordic Cooperation, 2017). To put the strategy into practice, governments in the Nordic member countries are simultaneously mobilizing investments at the county and municipality levels to boost bioeconomy to promote rural growth and create new jobs. The Nordic countries have simultaneously designed research and innovation programmes, where business enterprises involvement is strongly encouraged.

In addition, the Nordic Bioeconomy Panel is pushing for fundamental changes to production systems, aspiring to eliminate waste entirely and avoid negative impacts on the climate and the environment (ibid.). The Panel, together with other agencies, has catalogued 25 successful cases of bioeconomy so far, and the lessons learnt from these cases were further used in preparing a joint Nordic strategy for bioeconomy.

Bioeconomy development in Japan

In 2008, the Cabinet Office in Japan released a national strategy called 'Drastic reform with effective and agile movements for biotechnology innovation in Japan' (Cabinet Office of Japan, 2008). This strategy proposed: (1) the creation of new industries through prompt implementation of excellent basic research results; (2) the promotion of public awareness of research and development of genetically modified crops to solve food problems; and (3) research and development on the use of biomass to solve environmental problems. These were set as the priority issues to be addressed (ibid.).

There was a long silence before the latest strategy was released in June 2019 (Cabinet Office of Japan, 2019). It was named 'Bio-strategy 2019' and the aim of this strategy is to realize the most advanced bioeconomy society by 2030. The three elements that will enable the strategy's implementation include: (1) creating an environment where biotechnology can be pursued, addressing ethical, legislative and societal problems; (2) promotion of transdisciplinary and international communities and cooperation, which attract human resources and funding for research and innovations; and (3) bio and digital data to be recorded and integrated to form a database. Five goals have been set to realize this plan for society:

1 targeting the proper market;
2 integration of bio and digital data;
3 formation of international research hubs;
4 reinforcement of knowledge rights and genetic resource protection;
5 addressing ethical, legal and social implications.

However, the new strategy is rather inclined towards biotechnology and related innovations and lacks a proper notion of rural community and ecosystems.

There are several promising technology-dependent innovative bioeconomy cases emerging in Japan, for example, complete aquaculture of high-value fish, such as tuna (Aquaculture Research Institute, 2019), controlled environment sericulture (Silk on Valley, 2019) and the Euglenophyceae factory (Euglena, 2019), which are already commercialized. A comprehensive national strategy to link individual technology cases to rural welfare and ecosystem management is awaited.

Bioeconomy in South and South-East Asia

The concept of bioeconomy is rather new to the region. So far, in the region, only Malaysia and Thailand have a dedicated bioeconomy strategy, whereas others such as India, China, Sri Lanka, Indonesia and South Korea have relevant policies that only mention bioeconomy-related activities with emphasis on biotechnology and bioenergy (Lippe, 2016). The strategy's implementation in most of these countries is undertaken by the Ministry of Science and Technology, whereas departments dealing with the bio-based resources are under other ministries. This could become a constraint for cross-sectoral cooperation in bioeconomy planning and implementation in the future, as the respective ministries may not be willing to give away their power and jurisdiction to others. Another foreseen problem is the lack of skilled resources and adequate investments that will hinder the development of bioeconomy (Mena, 2001).

In 2012, the Malaysian Government introduced the Bioeconomy Transformation Programme (BTP) that provided a platform for the private sector to channel and maximize commercial opportunities in bio-based industries (FAO, 2018). This was followed by the Bioeconomy Community Development Programme (BCDP) in 2014 and the National Biomass Strategy, 2020, expanding on the bioeconomy value chain to optimize the use of biomass resources. In fact, Malaysia was one of the first countries in the region to develop a 'Bioeconomy Contribution Index' in the region that included five criteria – value addition, exports, investments, employment and productivity. However, environmental sustainability was not included in the criteria, unlike in the EU, which is a potential risk, if not addressed. There is also a risk that focus on bioenergy production, if combined with oil palm expansion, has the potential to cause serious problems such as deforestation and biodiversity loss, which are already a global challenge.

Another country that is keen on promoting bioeconomy in the region is Thailand that developed its national 'bioeconomy scheme' in 2017. In fact, bioeconomy was included in the government's phased economic development plan, *The Thailand 4.0*. Under this Plan, the Thai government is expected to mobilize 400 billion Thai Bahts as an investment in three phases (starting from 2018–2026) to support bioeconomy development (SEI, 2018). The focus will be on value-adding and diversification of products in the agriculture sector that supports the livelihood of millions of people in the rural areas. Thus, bioeconomy development within agriculture and biotechnology to provide food for the future will be a short-term goal, whereas development of biofuels and biochemicals will be the goal for the long term, as per the Plan. Overall, bioeconomy initiatives are expected to increase the efficiency in input and output use, thereby reducing waste.

Increasing demands from the bioetech and the bioenergy sector can compete with land and water resources that are necessary to meet food security. This is supported by the current policy that focuses on energy perspectives and neglects interdependencies between energy, water and food (Lippe, 2016). Stakeholders, especially farmers groups, were not involved in developing the scheme, and this may prevent the farmers growing the major crops, such as rice, sugar cane, rubber and cassava in the country, from benefiting from the bioeconomy scheme. Waramit (2012) emphasized the need for adequate investment in research and development, which is quite low in Thailand right now, in addition to a regulatory framework that is implementable, incentives for investment, and capacity-building of human resources to ensure the success of the sustainable bioeconomy in Thailand.

Status of bioeconomy in Africa

As Adam Smith said, farmers never get rich. But this is particularly true in Africa, where farmers not only face the traditional price squeeze, despite good harvests, but also challenges due to market structures and regulations. Increased international trade and global prices don't seem to work too well for agricultural goods (for obvious reasons) and seem opposed to ideas central to sustainability. So, the intriguing question is: *how will African farmers be persuaded to adopt new technologies if gains continue to be realized downstream?*

According to FAO (2018), bioeconomy is now being adopted by several low- and middle-income countries as part of their new vision for development. If there is any region that needs bioeconomy, it is Africa that is witnessing rapid changes in trade, population growth and urbanization. Business as usual can no longer meet the increasing demand for jobs for the youth and fuel economic growth in Africa. As biomass resources form a significant part of the economy, value-added processes must be combined with primary production to increase business and market opportunities. In this context, bioeconomy can provide a potential approach to add value to primary production, especially in sectors such as agriculture, forestry, aquaculture and livestock in Africa (ICIPE, 2018). There are others who advocate an 'inclusive bioeconomy' to reduce poverty in Africa and ensure social, economic and environmental sustainability (ACTS, 2017). Poku et al. (2018) observed that a general problem is the lack of coordination between farmers, processors and industrial end-users along the cassava value chain in their study in Ghana. A study by InnovAfrica (2018) in six African countries has shown weak links between maize-legume value chain actors at different stages, which needs to be strengthened to improve competitiveness and income opportunities. In general, strong policy support and investments are required to foster the emerging bioeconomy in Africa.

There are a very few initiatives to promote bioeconomy in Africa as of today. Hence, there is a need to develop a region-focused bioeconomy strategy supported by evidence-based policy, institutional arrangements and investments to promote innovative research and build capacity. Programmes such as BioInnovate mandate help to facilitate collaboration between scientists, policy-makers and industry partners, to link bioscience research ideas and technologies directly to business and the market (ICIPE, 2018). Agencies such as INBAR (2019) help in harnessing productivity and efficiency gains in the whole biomass-producing, processing, and trading system in Africa.

On the African continent, South Africa was one of first countries to develop a dedicated National Bioeconomy Strategy, encompassing agriculture, industrial and environmental bio-innovation and health with indigenous knowledge systems (IKS) as an important cross-cutting issue (DST, 2013; Green Growth, 2014). The South African government views bioeconomy as a significant contributor to the economy by 2030 through new industries that can generate bio-based services, products and innovations. The Bio-economy Strategy gains its legitimacy since it is aligned with the National Development Plan of South Africa, which considers science, technology and innovation key to the South African developmental agenda (DST, 2013). In addition, the relevant ministries, including Trade and Industry, Health, Agriculture, Forestry, Fisheries and Environmental Affairs, are involved in developing the research and development agenda. Other African nations can learn from South Africa as they try to develop their strategy for bioeconomy with an inclusive approach.

Other countries, including Uganda, Kenya, Tanzania, Mozambique in Eastern Africa, and Mali, Senegal and Nigeria in the West, have started to include bioeconomy-related initiatives in their national strategies.

Opportunities that bioeconomy can offer to address global challenges

While the majority of the developed countries can secure food and have economic power to import energy, and materials, developing countries in Asia and Africa are challenged by deficiencies. If bioeconomy is to create a new opportunity for development, such countries should have yet unexploited biomass or large waste streams of biomass.

There are four main domains of bioeconomy: energy, materials, food and medicine. Energy and materials demand high quantities and homogeneity of materials for better productivity. They are exposed to strong intersectional or inter-regional competition. Foods have higher value and can be differentiated with advances in technology. Medicine has the highest value per weight of product and it demands advanced technology for production. The potential for new medicine relies very much on the biological diversity of the nature. There is no single measure to show the potential or limitation of bioeconomy. Figure 1.2 shows the average primary productivity of each country together with the rate of undernourishment.

Bioeconomy has great potential in areas where net primary production (NPP) is high. Primary production is the synthesis of organic compounds from atmospheric or aqueous carbon dioxide that occurs through the process of photosynthesis. Net primary production accounts for loss by respiration. It is high in tropical zones where natural vegetation is rich with abundant precipitation. Because of this regenerative ability, many parts in

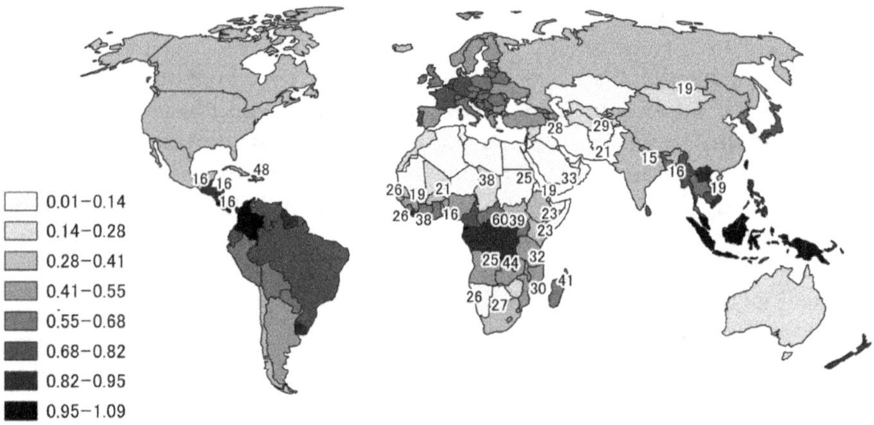

Figure 1.2 Average net primary productivity of countries (kg C m^{-2} year^{-1}).

Note
The numbers on countries are percentages of undernourishment if more than one source.

this zone have been converted from tropical forest rich in biodiversity to plantations. Contrarily, there is less opportunity for bioeconomy where availability of unused biomass is low. There can be a higher chance of food-material conflict if bioeconomy is introduced to countries low in food security. Most of the countries with high rates of undernourishment face political instability. In such countries, the opportunity for bioeconomy could be low.

If properly planned, integrated and implemented, bioeconomy is expected to contribute to new jobs, growth in the economy, optimal use of bio-resources, increased efficiency of input used and a reduction in greenhouse gases (EPRS, 2017; Golden and Handfield, 2014). This is the premise on which bioeconomy development should be based and followed in order to generate both short-term and long-term effects on the economy, society and the environment. It also opens doors to link the regional bioeconomies into the global bioeconomy development where countries could contribute to the implementation of the 2030 Agenda for Sustainable Development of the UN and its Sustainable Development Goals (SDGs) (SEI, 2018). Further, innovations and technology development in bioeconomy offer enormous opportunities for the private sector to innovate products that meet the global challenges, such as climate change, urbanization and resource scarcity (UPM, 2018).

The optimizing of biomass use comes with the 'cascading-principle use' in bioeconomy, where not only the main products, grain or timber, but also the by-products are used for multiple purposes. For example, to produce biofuels, bioplastics are produced that can promote green growth. This could also help in reducing pressure on land, and thereby reducing GHGs. In most of the developing countries, biomass production and use of resources are not optimal. Long-term strategies, political will and stability in policies are needed to exploit the opportunities bioeconomy can offer. It may help if the barriers to investments are removed, adequate capacity is developed, tax relief is provided, and incentives given where necessary. In Africa, where the potential for biomass production is high, governments need to think how to exploit the opportunity, access the technologies and growing markets and integrate the smallholders into the value chain enhancement (SEI, 2018).

Bio-based value chains typically depend on a multidisciplinary approach, which gives rise to new cooperation possibilities and alliances between sciences, sectors and agencies. Newly emerging bio-based knowledge hubs, as observed in Norway and other EU countries, with unique strengths are now extending their capabilities into new sectors (Innovation Norway, 2019). The three sectors, mainly, marine, agriculture and forestry, offer enormous opportunities for bioeconomy development, for example, the forest-based pulp industry redirecting the production towards micro- and nano-cellulosic fibres and composite materials. In the marine sector, scaling up marine processes and by-products, and the blue-green cross-sectorial opportunity, using nutrients leached from the land to the sea to grow seaweeds and harvest fish, and bioprospecting will be important for food security and sustainable development in the future. The principles of bioeconomy are relevant to mountain and coastal communities who have coped with the marginality and fragility of their ecosystems for centuries. As these ecosystems are now threatened by climate change, natural disasters, population growth and unsustainable management, bioeconomy can offer solutions to reverse these trends (*Bio-based News*, 2018).

A review by Johnson and Altman (2014) demonstrated the opportunities for rural development with specific reference to the implications of bioenergy and how it will transform the economies of rural regions. Teras et al. (2014) reported that bioeconomy is already proving to be an engine for providing jobs and economic growth in rural regions while reducing the impact on the environment. The large-scale impacts of bioeconomy development are possible only if properly scaled-up strategies and market access are developed with a systemic approach.

Constraints or risks with bioeconomy

According to Kleinschmit et al. (2017), bioeconomy in itself cannot be considered 'self-evidently sustainable'. Their study, reviewing bioeconomy discourses in Finland, Germany, France and the Netherlands, showed that the respective national bioeconomy strategies included environmental sustainability but mainly as a challenge rather than as an independent goal. Due to inherent innovation needs and uncertainties associated with bioeconomy, the chances of risks or failure are high (Golden and Handfield, 2014). Often the risks or uncertainties are due to biomass availability, climate and extreme weather, lack of suitable technologies and skilled manpower, social acceptance, political instability, market fluctuations and competition. Risk and uncertainty are linked, according to some, leading to negative impacts, whereas others view them as two different issues (Camilla et al., 2017). Within bioeconomy, it may be useful to consider their connection, given that biomass production is dependent on several environmental, technological and financial fundamentals.

A study by the TransNational Institute (2015) states that it is the powerful corporate sector lobby that is pushing the bioeconomy agenda, supported by the EU and G7 member states. This could lead to resistance from governments in developing countries. The complexity and interdependencies of SDGs can also pose constraints, as attempts to achieve one SDG can conflict with another. For example, SDG 2 (reducing hunger and ensuring food security) can conflict with SDG 16 (life on land) if adequate environmental considerations are not considered in increasing production. A multi-dimensional approach will thus be necessary to keep the balance, supported by innovative technologies, and legal and policy instruments. In the 1980s and 1990s, deeper rifts existed that reflected classical, opposing perspectives, related to fundamental views about human beings and technological progress, markets versus regulation. The 'green economy' approach largely wiped away these contradictions. It is therefore important for bioeconomists to learn from the past, take the multi-dimensional perspective into consideration and maintain a balance between the desired goals.

According to Hasenheit et al. (2016), bioeconomy encompasses a broad range of activities spread out along different value chains, both horizontal and vertical, that are interlinked, and managed by several actors who are connected in a multi-dimensional manner. As a result, the impacts of bioeconomy will not only be in the places where the biomass is produced, but also elsewhere, where the bio-based resources are processed and marketed. This implies that the impacts can affect communities and ecosystems beyond the borders, more so with global trade in bio-based resources rapidly increasing.

Technology-related risks

The basic requirements for biomass production to support a growing bioeconomy in any given situation are the availability of a sufficient quantity of produce, in a continuous and reliable supply, of good quality and sustainably produced (Lippe, 2016). This implies a good understanding of the competing uses and users, the need to find a balance between the potential and sustainable biomass production and promoting integrated approaches. Camilla et al. (2017) confirmed that the bio-based sector poses challenges due to significant feedstock availability constraints, limited technological data, and uncertain markets for the newly developed products. The study recommended that decision-making strategies must consider developing solid risk estimation procedures, easy-to-use tools for risk measurement and introduce mitigation measures as necessary. In addition, access to innovative technologies is limited, particularly, in developing countries, and there is a lack of willingness to share technologies and knowledge due to market competition.

The importance of knowledge, both traditional and scientific, in contributing to development since human evolution is undisputable. New research and technologies during the last 100 years have been successful in addressing several global challenges. At the same time new technologies can also pose risks if they are not properly used. Hence, low risk technologies that require minimum investments should be encouraged. Lack of appropriate technology in developing countries often forces them to compromise on the principles of sustainability that can lead to negative environmental impacts. Sharing knowledge and technology is one critical factor that needs to be addressed if the benefits of bioeconomy are to reach out to the socially deprived populations in developing countries. Some countries have the technology advantage, whereas others have the biomass resources. Genuine cooperation between different countries, sectors and stakeholders is therefore crucial to realize the full benefits of bioeconomy, where partnerships, transparency and knowledge-sharing should be an integral part (specified in SDG 17).

Environmental constraints

One of the arguments against bioeconomy is that the decrease in emissions is greatly over-estimated (TransNational Institute, 2015). The major concern is that the use and consumption of bio-based resources will in fact increase under the bioeconomy banner, without actually reducing the use of fossil-based resources. A number of critics argue that biofuels are produced at the cost of food security, livelihoods and taking land away from food crops in the Global South. If not planned and implemented properly, bioeconomy can lead to an increase in the use of biomass resources and higher GHG emissions than before with the overall increase in net production (European Parliament, 2017). This could increase competition between various sectors for limited resources, especially land and water.

Industrialization of bioeconomy that puts more emphasis on profits and business models and pays less attention to social and environmental objectives can create challenges for sustainable bioeconomy development.

The case of the oil palm sector in Thailand and Indonesia has shown that focusing only on economic goals can be detrimental to natural forests. This is a major risk when economic efficiency becomes the main criterion for bioeconomy development, ignoring the environmental externalities and other social costs (Zilberman et al., 2013). Hence, we need proper instruments in place such as payments for ecosystem services, green labelling and taxes. Economic efficiency, at anything greater than the level of the firm, in fact, should take externalities into account.

Countries that are putting emphasis on economic and business development pose a risk to natural resources. In the long run, this could lead to negative impacts on the natural resources if there is no balanced approach to the social, economic and environmental goals. Sustainable development seems to be the overarching concept included only in the political economy discourses and not put into practice in these countries. Thus, bioeconomy could become another paradigm like the Green Revolution of the 1970s, if environmental sustainability is not prioritized. Measurable sustainability indicators and a monitoring framework at all levels must be developed for bioeconomy programmes.

There are already standard indicators that can be used to measure the contribution of bioeconomy to SDGs. For example, in the Nordic countries, the contribution of bioeconomy projects is evaluated based on five criteria: the sustainable use of natural resources, technology innovation, environmental benefits, societal benefits and business model innovation.

Economic constraints/risks

Bioeconomy requires continuous investments to support research and innovation, and a good regulatory framework, to ensure its implementation and monitoring (Zilberman et al., 2013). This is a major constraint in developing countries where finances are limited. Given other national priorities, low- and middle-income countries may not be able to sustain the investments or subsidies necessary in bioeconomy. To mobilize funds, politicians need to be convinced about the impacts that bioeconomy can generate. Scientists working with bioeconomy must be prepared, having the necessary data and tools to measure the effects. Wesseler and von Braun (2017) report that measuring bioeconomy impacts is a challenge, as it cuts across several disciplines and sectors. However, this can be addressed through systematic data collection and analysis, developing baseline indicators and economic tools for measurement. A study by Muizniece et al. (2016) in Latvia showed that the bio-based sector is labour-intensive with low capital investments. Hence, there is a lack of political will to promote bioeconomy. The general assumption is that introducing new technologies to promote bioeconomy might lead to loss of jobs which could result in social unrest and risk to the communities in rural areas engaged in bio-based resources.

Governance and policy constraints

A comprehensive study of bioeconomy in 40 countries by Dietz et al. (2018) showed that an effective governance framework is very important for the development of a sustainable bioeconomy. However, governance in this context has two sides, one that is 'enabling', providing incentives and measures to promote bioeconomy, and the other restrictive, to control or regulate the use of bio-resources (ibid.). Both are necessary to ensure sustainability, but at the same time they are politically challenging. How countries will address this challenge, and implement the enabling or restrictive measures will determine the success of bioeconomy to a large extent.

A continuous societal discourse on how we can reconcile economic growth and environmental sustainability in future will be necessary to ensure the success of bioeconomy (Georg, 2018). At the same time, it is necessary to have proper policy and legal instruments that are implementable, as well as cooperation between various ministries, stakeholders and international agencies when bioeconomy is promoted. The regulatory framework should be able to take care of the human and environmental externalities, but at the same time it should not be a hurdle for research and innovation (Zilberman et al., 2013). Chapter 11 in this book will address the importance of stakeholder cooperation and governance necessary in supporting sustainable bioeconomy.

There is also a lack of international consensus on which products and activities are included in bioeconomy (FAO, 2018). At the same time, it is not easy to systematically monitor bio-based production due to its scattered nature, which is often driven by small-scale producers, especially in developing countries where they constitute the majority. Consequently, its contribution to the national Gross Domestic Product (GDP) can be underestimated and may not be attractive to encourage politicians to give their support.

Within the EU, there is lack of consistency between bioenergy, forestry and solid waste policies. Lack of policy coherence can make it difficult to implement the *Bioeconomy Strategy* which cuts across sectors (European Parliament, 2017). The cross-sectoral nature of bioeconomy demands that all sectors must be engaged from the beginning. This must start with the establishment of an inter-ministerial body involving all the relevant ministries in planning development and implementation of a bioeconomy strategy. Opportunities for stakeholder and public engagement in the governance of the bioeconomy actually occur at the regional level (Hasenheit et al., 2016). Hence, adequate resources should be made available at the regional level to establish a platform for engagement. International cooperation in bioeconomy research and development must be improved through joint programmes in research, and platforms for sharing available knowledge.

Bioeconomy and global policies: climate change and ecosystem services

The United Nations Framework Convention on Climate Change currently addresses the improvement of material and energy efficiency as major mitigation measures (UNFCCC Secretariat, 2018). New technical innovation and business models which realize CE and waste-to-energy are now being promoted. It urges countries to direct public support to facilitate private financing of such technologies. At the same time, data on waste quantities and characterization needs to be improved. The convention also notes that nature-based solutions and ecosystem-based adaptations are important. Bioeconomy, creating new business models along the value chain of biomass, is basically in line with the context of the convention. In terms of adaptation, the convention addresses poverty and gender as the key obstacles to the development of the most vulnerable people.

The mission of the Convention on Biological Diversity is to protect biodiversity at three levels, namely, genetic, species and eco-system (UNCBD, 2014). Expansion of human activity, linked with industrial agriculture, is seen as the primary cause of loss of biological diversity according to the most recent report (IPBES, 2019). The Intergovernmental Science-Policy Platform on Biodiversity and Ecosystem Services (IPBES) reported that out of 1.75 million species that have been described and formally named, a million species are facing the danger of extinction (ibid.). Since industry, agriculture, forestry and fishery are all responsible for rapid land use change and thus the loss of biodiversity, bioeconomy could be another risk to enhance the loss if wrongly driven.

The United Nations Convention to Combat Desertification currently has set five strategic objectives to counteract land degradation and to increase the resilience of the affected populations (UNCCD, 2019). Increased occurrence of drought is a major threat to these actions. These affected areas have sparse vegetation, limited water resources and nutrient-poor soils, which limit the implementation of bioeconomy in general. At the same time, bioeconomy can play an important role here to improve the value chain of biomass which is often weak in these areas.

Justification and rationale for the book

Bioeconomy is relatively a new area of development. Research on bioeconomy and the literature related to bioeconomy are limited today, especially outside the EU. This book, therefore, attempts to carry out a comprehensive literature review and analyse the different perspectives of bioeconomy, the opportunities it can generate, its barriers and constraints, and the measures required to realize the full potential of bioeconomy for the benefit of society. The various chapters in the book attempt to emphasize the link between a sustainable bioeconomy approach and the potential to achieve SDGs. This cross-cutting interdisciplinary book is targeted towards scientists, business managers, development agencies and academics dealing with international sustainable development, climate change, food security, or development economics as well as to policy-makers in governments and international agencies engaged with the SDGs. It will also be a useful reference for professionals in non-governmental organizations who are interested in development issues in general.

The structure of the book

In Chapter 1, the book provides a comprehensive review of the different bioeconomy perspectives, opportunities, constraints and recommendations for developing a mature and sustainable bioeconomy. This is followed by several case studies (land- and water-based) from different geographical regions, including the EU, Asia and Africa, covering agriculture, forestry, fisheries and bioenergy (Chapters 2–10). The book chapters are based on the experiences of several authors working with bioeconomy in the field. Chapter 11 emphasizes the need for stakeholder engagement, participative governance and transparency in promoting bioeconomy. The final chapter, Chapter 12 summarizes the key messages from the different chapters and suggests recommendations for sustainable bioeconomy development.

Conclusion

The literature review has shown that bioeconomy is rather a new concept, at times its goals are conflicting, and its objectives are in opposition to each other. Experience with bioeconomy is limited, and the concept only started to gain importance in the EU in 2009. Outside the EU, especially in Asia and Africa, bioeconomy is still not a mainstream topic. In the developed world, environmental sustainability is one of the main priorities while pursuing bioeconomy, while it is mainly for economic reasons that developing countries are promoting bioeconomy. Hence, the lack of a common bioeconomy agenda and understanding across the globe will be a main constraint to achieving SDGs. As the impacts of bioeconomy spread beyond country borders, a common agenda is necessary to keep a balance between the economic, environmental and social objectives. To ensure sustainable bioeconomy development, we need balanced goals, targeted research and innovation, technologies that are affordable and low risk in nature, secured investments, genuine stakeholder engagement and political commitment to put these conditions in place.

References

ACTS (2017) 'Inclusive bioeconomy'. Available at: www.acts-net.org/bioeconomy

Aquaculture Research Institute (2019) 'Completely farm-raised bluefin tuna'. Available at: www.flku.jp/english/aquaculture/tuna/index.html

Beddington, J., Asaduzzaman, M., Clark, M., Fernandez, A., Guillou, M., Jahn, M., Erda, L., et al. (2012) 'Achieving food security in the face of climate change: Final report from the Commission on Sustainable Agriculture and Climate Change'. Available at: https://cgspace.cgiar.org/bitstream/handle/10568/35589/climate_food_commission-final-mar2012.pdf

Bell, J., Paula, L., Dodd, T., Németh, S., Nanou, C., Mega, V. and Campos, P. (2017) 'EU ambition to build the world's leading bioeconomy: uncertain times demand innovative and sustainable solutions', *New Biotechnology*, vol. 40, pp. 25–30.

Bio-based News (2018) 'Bioeconomy in the mountains: an opportunity for local development'. Available at: http://news.bio-based.eu/bioeconomy-in-the-mountains-an-opportunity-for-local-development/

Cabinet Office of Japan (2008) 'Government-private sector joint promotion congress on biotechnology strategy'. Available at: www8.cao.go.jp/cstp/bt.html

Cabinet Office of Japan (2019) 'Bio-strategy 2019'. Available at: www.kantei.go.jp/jp/singi/tougou-innovation/pdf/biosenryaku2019.pdf

Calestous, J. (2016) 'Africa's bioeconomy'. Available at: https://qz.com/africa/860598/africas-bioeconomy-will-be-as-transformative-as-the-mobile-phone-and-digital-technology/

Camilla, A.A., Gonzalez, R. Kelley, S. and Jameel, H. (2017) 'Risk management consideration in the bioeconomy'. Available at: www.fs.usda.gov/treesearch/pubs/55507

Carus, M. and Dammer, L. (2018) 'The Circular Bioeconomy: Concepts, opportunities and limitations'. Available at: http://bio-based.eu/download/?did=120804&file=0

Dietz, T., Börner, J., Förster, J.J. and von Braun, J. (2018) 'Governance of the bioeconomy: A global comparative study of bioeconomy strategies', *Sustainability*, vol. 10, pp. 31–90.

DST (Department of Science and Technology) (2013) 'Launch of South Africa's bioeconomy strategy'. Available at: www.dst.gov.za/index.php/media-room/communiques/801-media-release-launch-of-south-africas-bio-economy-strategy

Easterly, W. (2015) 'The SDGs should stand for senseless, dreamy, garbled'. Available at: https://foreignpolicy.com/2015/09/28/the-sdgs-are-utopian-and-worthless-mdgs-development-rise-of-the-rest/

EC (European Commission) (2012) 'The Bioeconomy Strategy'. Available at: https://publications.europa.eu/en/publication-detail/-/publication/1f0d8515-8dc0-4435-ba53-9570e47dbd51

EC (European Commission) (2013) 'H2020 bioeconomy'. Available at: https://ec.europa.eu/programmes/horizon2020/en/h2020-section/bioeconomy

EC (European Commission) (2015) *Communication from the Commission to the European Parliament, the Council, the European Economic and Social Committee and the Committee of the Regions. Closing the Loop: An EU Action Plan for the Circular Economy*. Brussels: European Commission.

EC (European Commission) (2018) 'A new bioeconomy strategy for a sustainable Europe'. Available at: https://ec.europa.eu/commission/news/new-bioeconomy-strategy-sustainable-europe-2018-oct-11-0_en

El-Chichakli, B, von Braun, J., Lang, C., Barben, D. and Philp, J. (2016) 'Policy: Five cornerstones of a global bioeconomy'. Available at: www.nature.com/news/policy-five-cornerstones-of-a-global-bioeconomy-1.20228

Ellen MacArthur Foundation (2013) *Towards the Circular Economy*. Vol. 1, *Economic and Business Rationale for an Accelerated Transition*. Available at: www.ellenmacarthurfoundation.org/publications/towards-the-circular-economy-vol-1-an-economic-and-business-rationale-for-an-accelerated-transition

EPRS (European Parliamentary Research Service) (2017) 'Bioeconomy: Challenges and opportunities'. Available at: www.suschem.org/newsroom/bioeconomy-challenges-and-opportunities

Euglena (2019) 'Euglena projects'. Available at: www.euglena.jp/

European Parliament (2017) 'Bioeconomy: Challenges and opportunities'. Available at: www.europarl.europa.eu/thinktank/en/document.html?reference=EPRS_BRI(2017)595890

FAO (Food and Agriculture Organization) (2016a) *The Future of Food and Agriculture: Trends and Challenges*. Rome: FAO.

FAO (Food and Agriculture Organization) (2016b). *How Sustainability Is Addressed in Official Bioeconomy Strategies at International, National and Regional Levels: An Overview*. Rome: FAO.

FAO (Food and Agriculture Organization) (2018) 'Assessing the contribution of bioeconomy to countries' economies'. Available at: www.fao.org/3/I9580EN/i9580en.pdf

FAO (Food and Agriculture Organization) (2019) 'FAOSTAT'. Available at: www.fao.org/faostat/en/#home

Foley, J.A., Prentice, I.C., Ramankutty, N., Levis. S., Pollard, D., Sitch, S. and Haxeltine A. (1996) 'An integrated biosphere model of land surface processes, terrestrial carbon balance and vegetation dynamics', *Global Biogeochemical Cycles*, vol. 10, pp. 603–628.

Georg, S. (2018) 'What kind of innovation policy does bioeconomy need?', *New Biotechnology*, vol. 40, pp. 82–86.

Golden, J.S. and Handfield, R.B. (2014) *Why Biobased? Opportunities in the Emerging Bioeconomy. USDA Biopreferred Program*. Washington, DC: USDA.

Green Growth (2014) 'South Africa: The Bioeconomy Strategy'. Available at: www.greengrowthknowledge.org/national-documents/south-africa-bio-economy-strategy

Hasenheit, M., Gerdes, H., Kiresiewa, Z. and Beekman, V. (2016) 'Summary report on the social, economic and environmental impacts of the bioeconomy'. Available at: www.bio-step.eu/fileadmin/BioSTEP/Bio_documents/BioSTEP_D2.2_Impacts_of_the_bioeconomy.pdf

ICIPE (2018) 'Time for bioeconomy in Africa'. Available at: www.icipe.org/news/time-bioeconomy-africa

INBAR (2019) 'Toward a sustainable bioeconomy in Africa'. Available at: www.inbar.int/toward-sustainable-bio-economy-africa/

International Energy Agency (2018) 'Statistics'. Available at: www.iea.org/statistics/

InnovAfrica (2018) 'Agricultural food value chain pathways'. Available at: www.innovafrica.eu/

Innovation Norway (2019) 'Bioeconomy'. Available at: www.innovasjonnorge.no/en/start-page/invest-in-norway/industries/bioeconomy/

IPBES (2019) *Global Assessment Report on Biodiversity and Ecosystem Services of the Intergovernmental Science-Policy Platform on Biodiversity and Ecosystem Services*. Bonn, Germany: IPBES Secretariat.

Johnson, T.G. and Altman, I. (2014) 'Rural development opportunities in the bioeconomy', *Biomass and Bioenergy*, vol. 63, pp. 341–344.

Kalmykova, Y., Sadagopan, M. and Rosado, L. (2018) 'Circular economy: From review of theories and practices to development of implementation tools', *Resources, Conservation and Recycling*, vol. 135, pp. 190–201.

Kircher, M. (2014) 'The emerging bioeconomy: Industrial drivers, global impact, and international strategies', *Industrial Biotechnology*, vol. 10, no. 1, pp. 11–18.

Kleinschmit, D., Arts, B., Giurca, A., Mustalahti, I., Sergent, A. and Pülzl, H. (2017) 'Environmental concerns in political bioeconomy discourses', *International Forestry Review*, vol. 19, pp. 41–55.

Langeveld, J.W.A., Dixon, J. and Jaworski, J.F. (2010) 'Development perspectives of the biobased economy: A review'. Available at: www.researchgate.net/publication/220036536_Development_Perspectives_Of_The_Biobased_Economy_A_Review

Lippe, M. (2016) 'The emerging bioeconomies in South East Asia: New claims for biomass and land resources'. Available at: https://biomlandconf.files.wordpress.com/2016/02/biom-land-lippe.pdf

Mena, M. (2001) 'South East Asia: Gearing up for the bioeconomy wave'. Available at: www.frost.com/prod/servlet/market-insight-print.pag?docid=BNNN-5377MG

Muizniece, I., Timma, L., Blumberga, A. and Blumberga, D. (2016) 'The methodology for assessment of bioeconomy efficiency', *Energy Procedia*, vol. 95, pp. 482–486.

Nagothu, U.S. (2015) 'The future of food security: Summary and recommendations'. In U.S. Nagothu (ed.) *Food Security and Development: Country Case Studies*. London: Routledge.

Nagothu, U.S., Bloem, E. and Andrew, B. (2018) 'Agricultural development and sustainable intensification: Technology and policy innovations'. In U.S. Nagothu (ed.) *Agricultural Development and Sustainable Intensification: Technology and Policy Challenges in the Face of Climate Change*. London: Routledge, pp. 1–22.

Nordic Cooperation (2017) '25 Nordic bioeconomy cases for sustainable change'. Available at: www.norden.org/en/news/25-nordic-bioeconomy-cases-sustainable-change

Nordic Cooperation (2018) 'Nordic bioeconomy programme'. Available at: www.norden.org/en/bioeconomy

OECD (Organisation for Economic Cooperation and Development) (2009) *The Bioeconomy to 2030: Designing a Policy Agenda, Main Findings and Policy Conclusions*. Paris: OECD.

Pfau, S., Hagens, J., Dankbaar, B., and Smits, A. (2014) 'Visions of sustainability in bioeconomy research', *Sustainability*, vol. 6, pp. 1222–1249.

Poku, A., Birner, R. and Gupta, S. (2018) 'Is Africa ready to develop a competitive bioeconomy? The case of the cassava value web in Ghana', *Journal of Cleaner Production*, vol. 1, pp. 134–147.

Scarlat, N., Dallemand, J., Monforti-Ferrario, F. and Viorel, N. (2015) 'The role of biomass and bioenergy in a future bioeconomy: Policies and facts', *Environmental Development*, vol. 15, pp. 3–34.

SEI (Stockholm Environmental Institute) (2018) 'Global Bioeconomy Summit 2018: Opportunities and challenges in Africa'. Available at: www.sei.org/perspectives/challenges-opportunities-bioeconomy-africa-insights-2018-global-bioeconomy-summit/

Silk on Valley (2019) 'Silk in Yamaga spins the future'. Available at: http://silk-on-valley-yamaga.jp/en/

Su, B., Heshmati, A., Geng, Y. and Yu, X. (2013) 'A review of the circular economy in China: Moving from rhetoric to implementation', *Journal of Clean Production*, vol. 42, pp. 215–227.

Teras, J., Lindberg, G., Johnsen, I.H.G., Perjo, L. and Giacometti, A. (2018) 'Bioeconomy in the Nordic region: Regional case studies', NORDREGIO Working Paper, vol. 4.

TransNational Institute (2015) 'The bioeconomy'. Available at: www.tni.org/files/publication-downloads/tni_primer_the_bioeconomy.pdf

United Nations (UN) (2015) 'Sustainable Development Goals'. Available at: https://sustainabledevelopment.un.org/?page=view&nr=164&type=230

UNCBD (2014) 'Global biodiversity outlook'. Available at: www.cbd.int/gbo/gbo4/publication/gbo4-en-hr.pdf

UNCCD (2019) 'Committee for the review of the implementation of the convention'. Available at: www.unccd.int/official-documents/cric17-georgetown-guyana-2019

UNFCCC (2015) 'Adoption of the Paris Agreement'. Available at: https://unfccc.int/resource/docs/2015/cop21/eng/l09.pdf

UNFCCC Secretariat (2018) 'Climate action now summary for policymakers 2018'. Available at: https://unfccc.int/resource/climateaction2020/spm/introduction/index.html

UPM (2018) 'UPM presents the limitless opportunities of bioeconomy at the Pul-Paper 2018 exhibition'. Available at: www.upm.com/about-us/for-media/releases/2018/05/upm-presents-the-limitless-opportunities-of-bioeconomy-at-the-pulpaper-2018-exhibition/

Waramit, N. (2012) 'Developing a bioeconomy in Thailand: International Society for South East Asia', *Asian Agricultural Sciences Journal*, vol. 18, no. 2, pp. 34–44.

Wesseler, J. and von Braun, J. (2017) 'Measuring the bioeconomy: Economics and policies', *Annual Review Resource Economics*, vol. 9, pp. 275–298.

Zilberman, D., Kima, E., Kirschnera, S.S. and Reeves, J. (2013) 'Technology and the future bioeconomy', *Agricultural Economics*, vol. 44, pp. 95–102.

2 Sustainable biomass production from oceans and the potential for circular bioeconomy

Trond Rafoss and Udaya Sekhar Nagothu

Introduction

The oceans are receiving increasing attention in recent years due to their potential to offer solutions to the problems of over-exploitation of terrestrial resources and how to feed a growing human population. Typically, stakeholders project figures of the oceans' coverage of more than 70 per cent of the global surface, but oceans provide as little as 17 per cent of the human protein intake and only 2 per cent of the calorie intake (FAO, 2018a; SAPEA, 2017). The report by the Food and Agricultural Organization (FAO) in 2018 clearly shows the potential of the oceans to meet future food demands. Already in 2011, the European Commission (EC) adopted a Communication on Blue Growth, outlining the potential of Europe's coasts, seas and oceans for both jobs and bioresources (EC, 2011a). The Communication also pointed to the major constraints that could hamper Blue Growth, which include inadequate research and knowledge gaps, disputed property rights over territorial waters and conflicts between nations over the claims to ocean resources. The EC report stated that, in addition to food, the ocean food production can clean up the waters and thereby restore degraded ecosystems (EU, 2015). Sustainable food production from the oceans can contribute directly to the UN Sustainable Development Goals (SDG), SDG 2 (reducing hunger) and SDG 14 (life below water). In the process, it is very important to reverse pollution, manage and sustainably use the marine resources well. The actions leading to SDG 14 call for increasing measures to protect sensitive coastal and marine ecosystems and to limit fishing of endangered species. At the same time, it emphasizes the need to ensure livelihoods and economic growth of the coastal communities in a sustainable manner (MFA, 2017). In our opinion, it is not easy to achieve SDG 14 if we continue the way we are managing marine resources today. In this chapter we argue that a different approach is needed: an approach where livelihoods are secured for future generations by sustainable use of ocean resources in a way that produces harvestable output and at the same time reverses pollution and eliminates environmental degradation, as well as moving resources back on land.

Increasing food production from the oceans can have associated risks if proper conditions are not created and regulations are ignored or not in place. We are already witnessing serious problems with overfishing and depletion of fisheries in many regions across the globe. According to Tullis (2017), five out of six fishing areas worldwide have gone beyond the limits of sustainability. Similar risks may arise with unsustainable ocean food production, resulting in wholesale shifts in the composition, structure and function of the food web systems and this will ultimately affect the goods and ecosystem services provided by oceans, if we do not follow a sustainable path. The risks could lead to further pollution of oceans, the destruction of ecosystems, such as coral reefs, mangroves, eelgrass meadows and kelp forests, and the extinction of marine species. In addition, research and knowledge on the macro-level, i.e. ecological stoichiometry, are limited. Therefore, the first reaction from scientists to suggestions to increase ocean biomass production by cultivation of the ocean is their concerns about ecological impacts. On one hand, the current scientific basis of the potential that the oceans and seas have to increase biomass production for human consumption is limited. On the other hand, it is relatively well known that extreme amounts of plant nutrients, e.g. in the form of agricultural runoff and sewage, and CO_2 are injected into the ocean space. And the world is also well aware of the subsequent impacts, as many regions of the world are suffering the typical adverse effects of such anthropogenic releases leading to pollution of coastal waters in the form of eutrophication. What is beyond doubt is that the oceans play a key role in the world's climate system; hence, the way humans use marine resources will have a large impact on the environment and the future of human well-being.

Circular bioeconomy and ocean production systems

This chapter, therefore, explores the unique ocean opportunities to solve the grand global challenges, particularly, food security and climate change adaptation. The chapter will build on the assumption that ocean biomass cultivation has a wider scope than just the supply of food and energy to humans – it can provide a circular bioeconomy by: (1) providing food security; (2) recovering degraded ecosystems; and (3) recycling resources lost to the ocean back into terrestrial food webs. The ocean is already absorbing a major burden from human ruthless behaviour on Earth. Therefore, this chapter also warns that the conventional approach of over-exploitation of natural resources and pollution of the environment cannot continue. The conventional legal approach does not seem to be sufficient to secure a move towards sustainable development. Instead a new approach is required, cooperating with industry and other stakeholders to create a circular bioeconomy approach that, as well as feeding the growing human population, can put us on track towards sustainable development by an ocean-land circular bioeconomy.

As well as offering the opportunities for biomass production and harvesting, the oceans have the unique feature that biomass can be produced without adding fertilizer or feed. Thus, cultivation and harvesting of biomass in the ocean or seas have the potential to cycle back nutrients transported from the land into the ocean, e.g. from agriculture, forests, sewage and other land use activities. When trying to understand the oceans' complexity and size, one can consider moving from the two-dimensional view to the three-dimensional view of the oceans' share of the biosphere. This brings the ocean part from a surface coverage of more than 70 per cent to an amazing ~99 per cent share of the living space on Earth and indicates the enormous potential it has to offer.

The chapter justifies why mainstream nature conservation approach needs a paradigm shift – a change from ecosystem service protection goals to ecosystem service production goals. In this context, the chapter will address the following important questions

1 Where and how does the growth in global food supply from the ocean happen?
2 What are the major land-ocean biogeochemical interactions, constraints and opportunities?
3 What are sustainable land-ocean solutions?

The main societal hindrances to 'blue growth' are addressed in this chapter, which could be useful for scientists and policy-makers. The coastal zone constitutes the main interface to the ocean solutions that require strong scientific insights to secure progress and avoid problems. The chapter also highlights how sustainable production systems and consumption patterns have the potential to contribute to meet several of the UN SDGs.

The chapter starts with an introduction to the ocean biomass production potential and presents the cautious approach we need to take to develop bioeconomy. This is followed by a section on the ocean food production system and the basis for different marine species that can be produced without adding external chemical inputs, for example, typically low-trophic species groups such as seaweeds and molluscs. The bioeconomy opportunities from oceans are discussed towards the later part of the chapter, including a brief overview of the conditions necessary for sustainable land-ocean solutions. Finally, some technical and policy recommendations for promoting sustainable ocean-based bioeconomy are provided.

Ocean biomass production potential

The UN Food and Agriculture Organization (FAO) reported in 2018 that global fish production in 2016 reached an all-time high of 171 million tonnes (FAO, 2018a). Of this, more than 150 million tonnes were consumed directly by humans, meaning that each one of us consumed on average 20.3 kgs of fish[1] in 2015. The FAO concluded that the sector is crucial in meeting the goal of a world without hunger and malnutrition, by pointing to the fact that the annual global growth in fish consumption since 1961 has been twice as high as population growth. The report further projects an expansion in world aquaculture production and that this growth, though it will be slower than in the past, is anticipated to fill the supply–demand gap for the period up to 2030 (ibid.). Without further critical analysis, one might think that human society is on track towards a food-secure world by increasing the supply of nutrient-rich food from the ocean that will reduce pressure on the over-stretched land resources. However, behind these promising and optimistic reports about the future role and opportunities in human food supply from the oceans, new and complex changes are taking place that deserve attention and action.

Apart from the developments in aquaculture based on feed, which have grown faster than that of unfed species, there is, however, still a steady growth in aquaculture that is not dependent on feed. Nearly 30 per cent of total farmed food fish are unfed, according to the FAO (ibid.), constituting 8.8 million tonnes of filter-feeding finfish and 15.6 million tonnes of invertebrates, mostly marine bivalve molluscs. The overall figures of growth in world aquaculture production, reaching 110 million tonnes in 2016, have shown a stronger growth in farming of feed-dependent aquatic animal species than for unfed animal species, and with the latter share decreasing gradually by 10 per cent, to around 30 per cent from 2000 to 2016. The modern feeding aquaculture system uses pelleted feeds produced by large feed companies. In 1990, the feedstuffs used for salmon farming contained around 90 per cent raw material of marine origin (i.e. fish oil and fish meal made from wild caught fish) (Ytrestøyl, Aas and Åsgård, 2015) with the remaining 10 per cent a plant-based component added as a binder to support the pelleting process. This is a sign that deserves critical analysis to determine the direction of future development in the ocean food production system.

Moreover, the statistics for wild caught fish provide good news about an increased share of the catch being used directly for human consumption, while the proportion processed for fish oil and meal is stable. The increased share for human consumption is explained by reduced loss and wastage between landing and consumption, although they still account for an estimated 27 per cent of the fish landed.

The above-mentioned facts of continued growth in aquaculture production, with stagnant but stable wild fish catches, but with an increasing share used for direct human consumption and no increase in processing for fish meal and oil, all beg the question: *where does the growth in feed supply for aquaculture come from?* The answer is that the ingredients are mostly coming from agriculture. For instance, the feed composition in Norwegian salmon farming changed dramatically from 1990 to 2013

(Figure 2.1), with an increase from about 10 per cent to nearly 70 per cent in the terrestrial share, i.e. plant protein, oil and starch, and a corresponding reduction in the marine component from 90 per cent to less than 30 per cent of fish oil and meal in the feed (ibid.).

In terms of sustainability, this change is positive as it moves farmed fish down the food chain to a lower trophic level. In the case of farmed Atlantic salmon, it has changed from being a 90 per cent carnivore to a 70 per cent herbivore. However, the change in feed composition casts doubt on whether such seafood really is food from the ocean. As a matter of fact, the links between mariculture and agriculture are strengthening, although in a different way from what would be best if the aim were that the ocean only should take a larger burden of the human food supply. What is certain, is that the current growth in demand for aquaculture feedstuffs further contributes to the increased pressure on land-based resources, which is not a positive sign. On the other hand, fish have more efficient feed conversion rates because they are poikilotherm organisms, i.e. not 'wasting' energy on maintaining a constant body temperature like typical agricultural livestock species.

At the time of writing this chapter, the Amazon rainforest was burning at a rate never seen before in modern history. The demand for soy-protein worldwide, not only to feed beef, milk, pork and poultry production, but also the fed aquaculture sector, such as from the Brazilian exports, partly

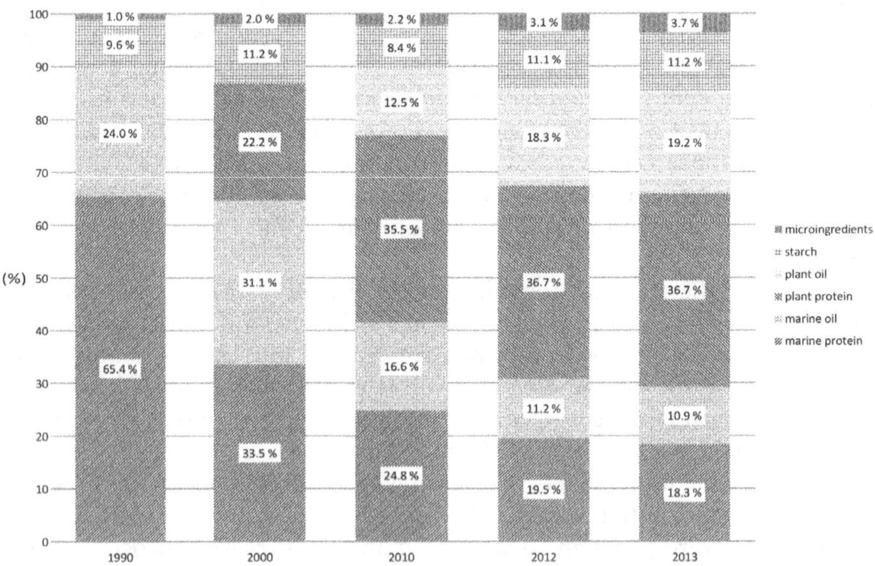

Figure 2.1 Ingredient sources (% of the feed in the Norwegian salmon) from 1990–2013.

Source: adapted from Ytrestøyl et al. (2015).

explains the reasons for man-made forest fires in the Amazon. The Amazon fires recently led to a critical shift in thinking by the Norwegian salmon farmers, who now realize that it is the total demand for soya feed from Brazil, whether it is certified with a non-deforestation label or not, that partly drives the Amazon deforestation. Consequently, 44 Norwegian salmon farmers, who make up the Salmon Group buying cooperative, have opted to remove Brazilian soy from their fish feed with immediate effect (*Fish Farming Expert*, 2019). Similarly, other concerned farmers and aquaculture companies the world over are now trying to find alternative sources for plant protein and oil (Ilaks, 2019). Although there is still a long way to go, it is an opportunity for the governments to work together with the private sector and the scientific community to develop evidence-based technologies and methods to develop alternative, sustainably produced biomass substitutes from the oceans. It is a positive sign when the industry itself draws this conclusion. At the consumers' end, there is also a growing demand, though not at the required pace, that the food produced be made in the 'right' way. The same may not hold for the poor communities who produce, harvest and live on a day-to-day basis. Any new sustainable systems or methods must therefore be matched to the context of their different socio-economic and environmental settings.

Marine-farmed aquatic plants

According to the FAO (2018b), the global production of farmed aquatic plants, dominated by seaweeds, has grown in output volume from 13.5 million tonnes in 1995 to more than 30 million tonnes in 2016. Most of this seaweed was of cultivated origin. Aquatic plants, as well as marine bivalves, have the ability to clean the organic waste particles from the water and at the same time extract dissolved plant nutrients from the water. These capabilities and their potential for solving global development challenges to food security and resource recycling are becoming increasingly recognized. The Norwegian research institute, SINTEF, recently blogged estimates of CO_2 capture by seaweed cultivation ranging from 3,000 to 1,500 tonnes of CO_2 per square kilometre, in high (20,000 tonnes of seaweed per square kilometre) to medium (10,000 tonnes per square kilometre) productive ocean waters respectively (SINTEF, 2019).

The Norwegian Institute for Water Research estimates that the Norwegian seaweed farming alone will reach 20 million tonnes by 2050, even though the current production is just a few hundred tonnes (NIVA, 2017). With the increase in seaweed cultivation, CO_2 fixation will significantly increase. Figure 2.2 illustrates the size of seaweed cultivation area (150 x 120 km) required to capture all CO_2 emissions for Norway, thus, demonstrating the enormous potential of seaweed to capture CO_2, and contribute to biomass

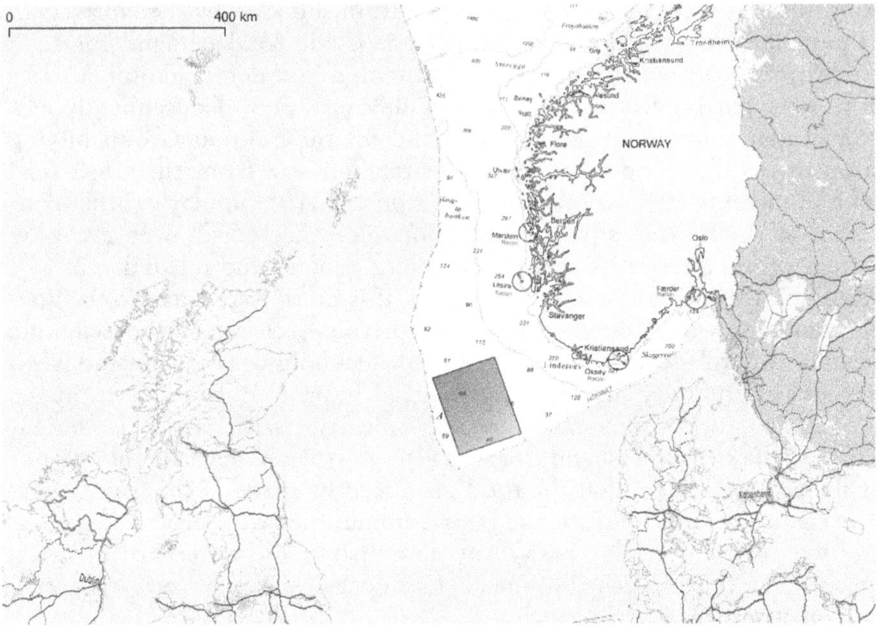

Figure 2.2 Seaweed cultivation area required to capture all Norwegian CO_2 emissions (54 million tonnes). Sea depths in the Norwegian economic zone added to inform about anchoring depths.

Source: based on map-data from Statkart.

production. About 99 per cent of the seaweed produced worldwide comes from eight Asian nations (Radulovich et al., 2015). These figures indicate that seaweed cultivation can be expanded in countries such as Norway, which has a long coastline and the capacity to invest and expand the seaweed sector. Norway's new strategy to promote the blue economy supports mariculture of aquatic plants, including seaweed and several fish species (MFA, 2017).

Historical use of seaweed in the western food tradition is sparse, with some exceptions, e.g. the use of dulse (*Palmaria palmate*) as a diet supplement for vitamins and other essential elements dates back at least to the Vikings. It is also well known that livestock can to some degree graze on seaweed growing in the tidal zone and on storm-beached kelps. Finally, the collection of storm-tossed kelp for distribution on agricultural fields as green manure is a tradition along the North Atlantic coast, especially before the invention of artificial fertilizers. In South-East Asia and the Far East, seaweed is widely used as food on a daily basis. Besides, seaweed-derived products can be put to multiple uses such as food, feed, medicine and other purposes.

Land-ocean interactions: the basis for ocean food production

Oceans are rich in elements essential for photosynthesis – the engine of bioeconomy. However, the limits to sunlight penetration into the water column, as well as stability in the ocean waters due to vertical stratification, cause regular depletion of nutrient resources in the euphotic zone (i.e. the upper part of the water column where sunlight is available). This paradox, in addition to sea ice coverage, explains why only about half of the global photosynthesis takes place in the ocean, despite its coverage of 70 per cent of the Earth's surface. The enormous bioeconomy potential of the oceans is revealed naturally in areas with so-called upwelling conditions (Kämpf and Chapman, 2016), where nutrient-rich deep water is brought to the surface that fuels intense production of marine biomass. The rich fisheries in areas like the Barents Sea and along the Chilean coast demonstrate the biological productivity potential of the ocean. Studies in the Barents Sea have shown that there are significant links between chlorophyll a and zooplankton biomass, as well as between net primary production and fish biomass, indicating bottom-up trophic interactions in this region (Dalpadado et al., 2014).

The bioeconomy potential, which is present in the ocean richness of inorganic resources like nitrogen, phosphorus and CO_2, inspires an idea of *sustainability by production*, where these essential elements for photosynthesis are used as inputs to cultivate ocean photosynthetic production for sustainable development, thus contributing towards a better environmental situation in the sea, and at the same time providing more food and energy for human consumption. Given the fact that these resources are further 'enriched' by anthropogenic emissions, currently resulting in ocean acidification and coastal eutrophication, this makes this idea more of a need, rather than an opportunity.

The stoichiometry of essential biogeochemical elements, such as nitrogen (N), phosphorus (P), carbon (C), silicon (Si) and iron (Fe), varies. These are determinants of ocean bioproduction, and which elements constitute the bottlenecks vary. The complexity is partly dealt with in Chapter 3 in this book, where the flow of iron from the land to the ocean and its impact are discussed in detail. The consequence of eutrophication is well documented in science and can be countered by biomass cultivation, while other processes are less understood, which need further investigations as we increase biomass production in the oceans. Land-use management and changes thereby in the coastal zone continuously affect the nutrient flows, which, in turn, can lead to more serious changes in carbonate chemistry than ocean acidification (Borges and Gypens, 2010). Here the concern is about the disruption of nutrient flows from the land, as we all contribute to sewage and pollution. Therefore, an understanding of carbonate chemistry and how it affects the rate of marine primary production and calcification becomes important. Lowe et al., in their (2019) study, report that local ecosystem processes modulate ocean acidification and they support the adoption of an ecosystem perspective to ocean acidification and multiple stressors in productive aquatic habitats. High photosynthetic uptake will lead to a decrease in CO_2 and a higher pH. Both eutrophication and acidification influence the carbon cycle.

In the case of eutrophication, governments in the Global North chose a different approach to solving the pollution problem. Especially during the 1990s and in early 2000, huge investments were made in cleaning and purification systems for public sewage. Such solutions are associated with the large costs related to investments and maintenance, and provide little or no productive output, unless the extracted resources can be used. Yet, such nutrient regulation policies have been mainly implemented in industrialized countries, while in emerging economies, in the Global South, eutrophication beyond a safe level continues to be a serious problem, causing negative impacts on the ocean ecosystems, particularly in the estuaries. Several studies have looked at the impacts of excess nutrients that can be highly damaging, leading to effects such as anoxia and hypoxia from eutrophication, algal blooms, dieback of seagrasses and corals, and reduced populations of fish and shellfish (Baden et al., 1990). Of concern are the chemical compounds and nutrients disrupting endocrine functions, namely, persistent organic pollutants (POPs), petroleum hydrocarbons and metallic compounds, such as those of mercury and cadmium (UNEP, 2001).

Nutrient regulations on the coastal zone place intense pressure on marine communities (Lovelock et al., 2009). Among the consequences seen is a shift towards phosphorus limitation of primary production due to nutrient regulation policies in agriculture, as described for the Belgian coastal zone. A study by Riemann et al. (2016), in Denmark, reported the effect of reduced nutrient inputs into the sea from the land by about ~50 per cent for nitrogen (N) and 56 per cent for phosphorus (P) since 1990, resulting in reduced phytoplankton biomass (measured as chlorophyll a concentration) and increased cover of macroalgae in deeper waters. Recently, the eelgrass meadows have also expanded into deeper waters, in response to improving water clarity. The discharge of nitrogen and phosphorus from agriculture and industrial sources is potentially the most damaging of all human activities on the oceans, in terms of both scale and consequences (UNEP, 2001).

In general, nutrients leached from land can be tapped and re-used through ocean-land bioeconomy and thus contribute to the sustainable production of edible and marketable products, provide fertilizers for agriculture and thus bring carbon from the ocean back on land for soil storage (for more details, see Chapter 3), and thus contribute to reduced climate change (SDG 13). Ocean-land interactions in nature are ambient as water, carbon and nitrogen cycles, while direct movement of biomass from the ocean onto land is sparse (e.g. throwing of eelgrass and seaweed on the shores by storms, migration of anadromic fish species). Examples of fish migrating from the ocean to freshwater in nature, in a way, bring biomass from the ocean on to the land by their spawning products, faeces and, to a large extent, also their body biomass, as especially anadromous fish species that have a high mortality during spawning or catadromous fish.

We hereby argue that a general approach to learning from biomass production processes ongoing in nature should mimic and upscale these processes by facilitating biomass production by cultivation. This is a better and more productive approach than costly 'clean-up' of the problems caused by anthropogenic activity.

Opportunities from the ocean-land bioeconomy

In an ocean-land bioeconomy, the potential is present to feed a growing human population and at the same time reduce pressure on terrestrial resources and reverse the ocean ecosystem degradation. Marine resources can be of both a food and non-food nature that have high market value. Sustainable production of these resources can offer enormous potential to sustain the livelihoods of millions of households, especially the poorest populations living on the coastal fringes in the Global South who are extremely vulnerable to climate change. A study by Nagothu and Ortiz (2007) showed that aquaculture in the Philippines was one of the main sources of income for most of the poorest households among the fishing communities who cultivated seaweed by using indigenous technology. The farmers were not organized and the yields were quite low due to lack of knowledge and production facilities. A similar situation prevails in most of the coastal regions in developing countries, where subsistence fishing is common, due to lack of technology, poor investments and no pro-poor policy.

In the West, ocean-land bioeconomy offers opportunities for large-scale aquaculture expansion with new technologies and market expansion. In addition, the developed nations have invested significantly in research and innovation, new strategies and follow-up action, which gives them an advantage to tap the potential of the oceans for biomass production. In the case of Norway, the recent government emphasis on the blue economy has opened doors for new entrepreneurs along the entire salmon value chain, in the research and innovation and knowledge exchange between stakeholders to boost bioeconomy (MFA, 2017). Similar policies and programmes have already been introduced in several EU countries to develop offshore aquaculture by establishing a technical infrastructure and data-gathering networks on offshore conditions and impacts (DAFF, 2007).

Farming of Atlantic salmon serves as an informative case for an analysis of general coast-land interaction and the challenges emerging from growth in the ocean bioeconomy in several aspects. Some of the challenges include feed production and composition; pest risks and area occupation (land/ocean), as well as interactions between these factors. Farming of Atlantic salmon originated in the early 1970s when the Norwegian animal breeding scientists Harald Skjervold and Trygve Gjedrem initiated classic animal breeding of salmonid fish. A major input and the challenge in fish farming were, and still are, the feed. As already mentioned, feed from marine sources was common during the 1980s and 1990s but is gradually reducing from 90 per cent in 1990 to less than 30 per cent in 2013 (see Figure 2.1). This indicates that ocean feed production is lagging, while the ocean-land interaction and production are increasing.

In fact, there are now alternatives available on the market. The first farmed Atlantic salmon fed by microbial (yeast) protein produced from brown seaweed and spruce wood was raised recently by Sharma et al. (2018). Even alternatives that can replace the last remaining portion of required marine ingredients in the fish feed will be on the market soon. The so-called Omega-3 fatty acids (Tocher et al., 2019) are essential to maintain aquaculture fish health, where eicosapentaenoic acid (EPA) and docosahexaenoic acid (DHA) are the two most important components. Omega-3 fatty acids are introduced in the first step in the marine food chain, as they are produced by ocean microalgae. These unicellular photosynthetic micro-organisms are then consumed by zooplankton, which again is eaten by pelagic fish. The latter is caught and processed into fish meal and oil. Throughout this food chain the essential fatty acids EPA and DHA, originally synthesized in the microalgae, are preserved, making the fish healthy for humans, although in recent years less so for farmed fish, as the content of Omega-3 fatty acids has dropped along with the reduced proportion of fish oil and meal in the feed.

Although the engine of bioeconomy is photosynthesis, industrial bioeconomy, however, may take surprising approaches. One such example is the recently established commercial production of oil from microalgae. As already mentioned, these micro-organisms are key producers of the healthy and nutritious Omega-3 fatty acids. However, the industry has found that the shortest path to commercialization is to grow the algae heterotrophic by feeding them sugar in the dark instead of autotrophic (photosynthetic) growth with carbon dioxide (CO_2), water (H_2O) and light as inputs.

Apparently, such a solution might look disappointing from a sustainability point of view as no new carbon is captured in the algal growth process and no new primary production takes place. However, when compared to the fishing industry's traditional hunt for these essential fatty acids through harvesting of pelagic fish, the carbon footprint is a different story. In the end, algae oil production, even if produced by microalgae in the dark, will be comparatively far better, as the input sugar has been produced by photosynthesis in sugar cane/sugar beet. It is relatively much more carbon-efficient than the fishing vessels trawling the ocean for pelagic fish to process for fish oil. In addition, feeding the microalgae directly to aquaculture fish jumps over at least one trophic level compared to the food web transition happening in nature.

Beal et al. (2018) presented a land-based model facility of 111 ha area that would be able to produce 2,750 tonnes of protein and 2,330 tonnes of algal oil per year, based on experiences from Hawaii. They estimated that 100 such facilities using marine water would replace all domestic production of fishmeal in Thailand. This method will be able to save 30 per cent of the world's fish catch by producing fishmeal and fish oil replacement products from marine microalgae. Marine water is a good medium for microalgae culture, avoiding the cost of nutrients and enhancing the productivity of lipids and other useful products in microalgae biomass. Also, it is used for fish feed and food. The potential of microalgae for biofuels is growing in recent years, since they are considered the perfect substitute for liquid fossil fuels with respect to cost, renewability, and environmental concerns (Khan, Shin and Kim, 2018). This might significantly increase the demand for seaweed-derived products in the coming years.

The recent developments in feed technology demonstrate the potential in both ocean feed production as well as further strengthening the land-ocean links. Scientists at the Norwegian University of Life Sciences raised and harvested the first Atlantic salmon (*Salmo salar*) fed on yeast protein, grown from substrate of mashed brown seaweed (*Saccharina latissima*) and spruce wood (*Picea abies*) logs, and this serves as an excellent demonstration of the innovation potential in marine-based bioeconomy, i.e. that primary production biomass both from a marine kelp forest and from a terrestrial spruce forest can be converted into high quality fish proteins for human consumption (Sharma et al., 2018).

Global mariculture production is at an all-time high now, of which more than 50 per cent of all products are cultured molluscs and aquatic plants, which extract their nutritional requirements directly from the ocean (FAO, 2018a). Extractive species is another collective name used for species that take their nutrition from the water, which have the potential for bioremediation, i.e. to improve water quality by removing organic matter and absorbing dissolved nutrient elements. As the production and processing of crustaceans and bivalves have increased, efficient use of their shells has become important, not only to maximize financial return, but also to address waste disposal problems because of their slow natural degradation rate. The multiple uses of chitosan produced from shrimp and crab shells make it a valuable resource and it is no longer seen as a waste product. It has a wide range of applications, for example, in water treatments, cosmetics and toiletries, food and beverages, agrochemicals and pharmaceuticals. The value addition of shells was possible due to new processing technologies that are able to extract the marketable components. The subsidiary processing units also provide jobs and additional income.

The area-wise potential for mariculture estimated by Gentry et al. (2017) showed that only 0.015 per cent of the global ocean area is required to produce all the current total landings from wild-capture fisheries. Technological advances during the last few decades in industries such as offshore oil and gas exploitation, fisheries, mariculture and more recently, offshore wind energy, allow marine large-scale operations to take place under harsh open ocean conditions. Innovations in sustainable mariculture on the technology front can unfold huge economic opportunities in the future with minimum negative impacts on the environment, if carefully regulated. This may not be easy to put into practice unless there is commitment from different actors and political action to cooperate. One benefit of plant and herbivore mariculture compared to agriculture is that it is independent of industrial fertilizers, feed, and large supplies of freshwater and there is a very large unrealized worldwide biological potential for producing more sustainably.

CO_2 binding by aquatic plant production and mitigation potential

Macroalgae species is an informal grouping of multicellular algae that can be seen without the use of a microscope. They attach themselves to hard surfaces with an organ named the holdfast. Macroalgae are commonly divided into three groups of brown , red and green algae.

As previously mentioned, recent calculations by SINTEF in Norway showed that the whole of Norway's CO_2 emissions (52.9 million tonnes) can be recaptured by photosynthesis through kelp cultivation on an area of 18,000 square kilometres of the ocean surface (SINTEF, 2019). That is 3,000 tonnes of CO_2 captured per square kilometre. Net removal of these CO_2 emissions either requires that this biomass is used as fuel to replace fossil fuels, used for non-decomposition purposes, e.g. building material, or it has to be deposited for sedimentation in deep sea valleys.

One main argument against such large-scale aquatic plant cultivation is that the photosynthetic growth will be limited by lack of plant nutrients. It is also criticized that such cultivation will reduce photosynthetic production in both the wild macroalgae, as well as microalgae, and thereby will lower the productivity of the natural ecosystem. A possible solution to this problem is to explore whether macroalgae can collect nutrients from the deep by using submergible seaweed farms. A research initiative is ongoing on the southern coast of Norway, where one of the authors is involved in exploring the feasibility of this solution through field experiments by submerging live sugar kelp to various depths and at various time periods during the night and measuring the nitrate concentration and uptake rates during the period of plant nutrient depletion in the top layer (Figure 2.3). In these experiments the eventual effect of pressure is also studied, by having one experimental treatment on shore in a land-based aquaculture facility collecting seawater from the various depths in parallel to the field experiment. Mortensen (2017) has already shown that the brown macroalgae take up nitrate from the water also under dark conditions.

Figure 2.3 Harvesting a test rig for cultivation of sugar kelp (*Saccharina latissima*) in Southern Norway by the company Arctic Seaweed.

Source: photo by Kim Kristensen.

Recently, a few innovative experiments have been carried out to explore the potential of utilizing plant nutrients from the deep. Mortensen (ibid.) conducted an experiment in Lysefjord, Norway, where freshwater from a hydropower station was pumped to the bottom of the fjord. This process is known as artificial fjord upwelling, whereby the lighter freshwater was released, creating an upward flow dragging the nutrient-rich water from the deep up to the surface. This is an approach where the nutrient-rich deep water is lifted by freshwater injected into the deep, which will force an upstream current, due to the difference in density in freshwater and in salty deep water.

Eutrophication of coastal waters is still an environmental problem in the North Sea, despite a successful campaign during the 1990s where municipal sewage treatment systems were built in all the countries surrounding the North Sea. At that time there were critical warnings from scientists about imbalances from sewage treatment as the methods in place are most efficient at P reduction (Lindahl, 2011). Similar concerns were expressed in Japan during the 1990s, where treatment facilities were 100 per cent efficient in their treatment of sewage water before disposal into the seas. Municipal authorities and fishing communities in Japan had to rethink and subsequently reduced the efficiency of the treatment systems to mutual agreed standards supported by evidence-based research.

In other regions, where sewage facilities were poor, it has led to a number of negative effects on the marine environment, such as poor visibility depth, oxygen-free bottoms and overgrowth of shallow bays – something that will require major environmental conservation efforts going forward. Large-scale mussel cultivation was seen as a realistic and cost-effective method of reducing the negative effects of eutrophication (ibid.). At the same time, this is a cycle-adapted food production that also returns nutrition from the sea to the land. Pollution of the oceans by untreated sewage and other solid waste, more recently, plastic has attracted the attention of the public and political leaders. The concern for the protection of the oceans is growing, but we need to do more in terms of action and investments to check both point and non-point sources of pollution that impact oceans.

A transformative change to manage the ocean food system sustainably

Man's dependence of ecosystem functioning for the provision of air, water, food and fibre is not disputed, but there is increasing concern about the ecosystem impacts of human activity.

Climate change is a strong signal that the role of man in the ecosystem needs a significant re-think. The latter approach calls for a different and more systematic approach where humans need to find their role as a part of the global ecosystem, instead of as an external body that can exploit and over-use ecosystem services without concern. Some scientific research tends to place humans outside the ecosystem and does not regard them as a part of the ecosystem. Typically, nature conservation science tends to adopt an approach where nature needs to be protected from man, instead of developing sustainable solutions for man as a species in the world ecosystem. In the authors' view, this lack of understanding permeates modern human society, while some cultures have a more holistic understanding.

Conditions necessary for sustainable ocean-based bioeconomy

Beyond providing man with food and fibre, bioeconomy could be the human tool to balance the impact on natural ecosystem functioning. This is relevant in the context of land-ocean interactions, where a major proportion of terrestrial bioeconomy inputs like N and P end up in the ocean simply because agricultural plants at their best absorbs around 70 per cent of the supplied fertilizer. Through marine aquaculture, these nutrients leached from land into oceans can be used, thus recycling the nutrients. As discussed in the previous section, large-scale cultivation of aquatic plants and filter feeders/herbivores like molluscs and harvesting (removal of biomass/organic matter from the ocean) demonstrate how sustainable bioeconomy from the ocean is possible.

In the search for future solutions to the grand global challenges for human society, world leaders increasingly are looking towards the ocean for solutions. SAPEA (2017) asked how more food and biomass can be obtained from the oceans in a way that does not deprive future generations of their benefits. When addressing this question, they found that they had to redefine the question as: 'How can more food and biomass be obtained from the oceans in a way that maximizes the benefits for future generations?' SAPEA (2017) put forward a number of options grouped into four main categories:

1 Improvements in management and increased use of wastes in the traditional capture fisheries.
2 Fishing of new wild species that are not, or only marginally, exploited today.
3 Mariculture of organisms that extract their nutrients directly from the water.
4 Mariculture of organisms that require feed.

We think these suggestions are both good and valid, but at the same time the underlying reasoning why we should obtain more food and biomass from the oceans should be emphasized further, especially, with respect to the potential environmental benefits and the role of man in the ecosystems.

We need genuine and stable partnerships at the local level, particularly regarding sharing responsibility for environmental stewardship among the public and private sectors and civil society to manage marine and coastal ecosystems sustainably (Nagothu, 2005). As emphasized in Chapter 11 in the book, partnerships engaging relevant stakeholders to cooperate are essential to ensure a sustainable bioeconomy. More so, when the boundaries in the oceans are disputed and spillover effects of pollution do not recognize national boundaries, international cooperation becomes an absolute necessity. In our view, public regulation of the aquaculture industry has in general been very weak, not only resulting in conflicts with other interests, but even causing catastrophe and collapse in some regions of the world due to the industry itself. Typically this results from over-use of the resource base, unsustainable intensification and immature technology. Key words for the causes of these problems are pollution and disease.

Thus, new principles have to be developed for the area-planning processes in coastal and near-shore sea areas. First and foremost, the underlying analysis of the resource base must consider ecological sustainability constraints and, second, economic sustainability factors. Furthermore, the planning process must consider other sectors' plans for utilization of marine resources to secure potential synergies, especially the energy sector, as it can provide the basis for common infrastructure, e.g. offshore windfarms, but also traditional industries' use of sea areas such as fisheries, tourism and maritime transport. If carried out properly, many conflicts can be avoided and there should be ample opportunities to achieve synergies with other industries, including fisheries.

When it comes to the natural resource basis for the various types of known and possibly unknown sea aquaculture productions, the biophysical resource base must be decisive. Former planning processes in the past have tended to take a two-dimensional approach, while it is the three-dimensional resource base that is of interest in a sea aquaculture context. That is, the water volumes to be made available and their properties regarding water quality, such as oxygen content, nutrient content, light supply and replacement rate (= current). An intelligent approach indicates that a more differentiated sea area management system must be developed where areas allocated to aquaculture do not do so on a general basis as before, but on what kinds of production are allowed and specified, based on the suitability of the water area and volume. Areas with good access to nutrients and light can be set aside for production of aquatic plants, while areas that are known for high content of microalgae (= food) and have a high-speed throughput/water current can be set aside for the cultivation of filtering organisms (e.g. shells and tunicates). Common to so-called extracting organisms, i.e. species that collect all their nutrition from the water, such as both aquatic plants/algae and filter feeders, is that they have great potential to purify the water mass (= improving water quality). For aquaculture of species that require feeding, water volumes must have sufficient recipient capacity, or faecal collection and uptake may be required through other biological productions (extracting species) and/or technical facilities.

In general, when assessing opportunities and suitability for land for aquaculture production, the extension of suitability assessments must not be limited to the three-dimensional space only. A fourth dimension should also be included, i.e. the time dimension. The seasonal suitability of different areas throughout the year is well known in most areas where aquaculture is operating. The winter season can offer anything from too cold water to icy conditions that can destroy equipment and infrastructure. The summer season can offer hot water conditions, poor oxygen and nutrition depletion. Similarly, there are challenges for various productions both in autumn and spring. Seasonal approaches to sea use for aquaculture can benefit from the rapid developments in land-based aquaculture, where production can take place on land during unfavourable seasons in the sea and/or reduce the production time in the sea in order to avoid biological seasonal problems at sea. The consideration of other seasonal industries may also have an impact on whether the establishment of new aquaculture productions will be appropriate. Seasonal restrictions on new permits can be introduced to utilize areas for production even if they are unsuitable for part of the year.

For example, in some Southern Norwegian waters, with a limited use, it is conceivable that more salmon farming can be allowed through an approach where only parts of the year are operative times. The release of large smolt in late summer or autumn when most tourists have left and early harvesting in the winter before parasite problems have emerged and the water has not become icy, may be a strategy that fits well with the general development of the salmon farming industry where the salmon are kept ever longer on land to shorten production time in the sea.

In general, when it comes to new production processes, and especially those with a high ecological sustainability potential, different types of stimulants may prove necessary to get started and gain profitability. Here, it is conceivable that conditions be set to produce certain quantities of different species in given areas, both of species with strong ecological sustainability, although economic sustainability has not yet been achieved, together with species with weak ecological sustainability but strong economic sustainability. In this way, the commercially profitable productions can force development (reduce production costs and improve technology) of the more ecologically sustainable production processes.

In many countries, previous and even currently applied principles in public area planning for coastal and open waters usage, within local or regional administrative borders, tend to be general and not very well developed. The consequence is often conflict. Sea aquaculture has developed from a business dominated by small and single farmer enterprises into an intensive and capital-demanding industry, driven by big companies noted on global stock exchanges. Future policies need to address and facilitate equal conditions allowing both the large players to develop sustainable production systems, as well as the small enterprises. The future ocean bioeconomy will need the innovation potential of both the big and the small actors.

Conclusion

Many questions remain unanswered in terms of how the wider ocean processes influence the living resources within the oceans' water. Sea surface temperature influences our weather systems, impacting both marine and land-based activities. In addition, human-induced activities seriously affect the ocean food production systems and bio-geochemical cycles. Investing in applied and basic research will reduce uncertainty and help in future planning of mariculture. At the same time, sharing research results and knowledge across countries and sectors will help in sustainably managing oceans, which is in the common interest of the global community. Finally, to conclude, the following are some important recommendations to support *sustainable ocean-based bioeconomy*:

- Introduce financial incentives to encourage innovations and sustainable production systems.
- Encourage seaweed and other low-trophic productions and fish farmers to form associations and common trading systems with guaranteed minimum prices and unlimited volumes on delivery permission.

- Initiate processing studies and monitoring systems to avoid eventual negative by-products from processing seaweed and other species.
- Establish common platforms to share data and knowledge on oceans.
- Fund long-term multidisciplinary research programmes through joint cooperation between agencies from different countries.
- Develop advanced area plans with a proper legal basis for both coastal and offshore waters based on a scientifically sound, biophysical resource base analysis for sustainable bioeconomy production in time and space.

Note

1 'Fish' includes fish, crustaceans, molluscs and other aquatic animals, but excludes aquatic mammals, reptiles, seaweeds and other aquatic plants.

References

Baden, S.P., Loo, L.O., Pihl, L. and Rosenberg. R. (1990) 'Effects of eutrophication on benthic communities including fish: Swedish west coast', *Ambio*, vol. 19, pp. 113–122.

Beal, C.M., Gerber, L.N., Thongrod, S., Phromkunthing, W., Kiron, V., Granados, J., Archibald, I., et al. (2018) 'Marine microalgae commercial production improves sustainability of global fisheries and aquaculture', *Nature*. Available at: www.nature.com/articles/s41598-018-33504-w

Borges, A.V. and Gypens, N. (2010) 'Carbonate chemistry in the coastal zone responds more strongly to eutrophication than ocean acidification'. Available at: www.researchgate.net/publication/237513215_Carbonate_chemistry_in_the_coastal_zone_responds_more_strongly_to_eutrophication_than_ocean_acidification

DAFF (Department of Agriculture, Fisheries and Food, Ireland) (2007) 'Ireland's marine sector and the bioeconomy'. Available at: www.gov.ie › policy-information › e551fa-the-bioeconomy

Dalpadado, P., Arrigo, K.R., Hiollo, S.S., Rey, F., Ingvaldsen, R.B., Sperfeld, E., van Dijken, G.L. et al. (2014) 'Productivity in the Barents Sea: Response to recent climate variability'. Available at: https://journals.plos.org/plosone/article?id=10.1371/journal.pone.0095273

EC (European Commission) (2011a) 'Innovation in the Blue Economy: realizing the potential of our seas and oceans for jobs and growth'. COM/2014/0254 final/2. Available at: https://eur-lex.europa.eu/legal-content/EN/TXT/?uri=celex:52014DC0254R(01)

EC (European Commission) (2011b) 'European Commission launches Blue Economy Report at European Maritime Day in Lisbon'. Available at: https://ec.europa.eu/maritimeaffairs/press/european-commission-launches-blue-economy-report-european-maritime-day-lisbon_en

EU (European Union) (2015) 'Sustainable agriculture, forestry and fisheries in the bioeconomy: A challenge for Europe'. Available at: http://ec.europa.eu/research/scar/pdf/feg4-draft-15_may_2015.pdf

FAO (Food and Agriculture Organization) (2018a) 'State of world fisheries and aquaculture 2018'. Available at: www.fao.org/3/I9540EN/i9540en.pdf

FAO (Food and Agriculture Organization) (2018b) *In Brief: The State of World Fisheries and Aquaculture: Meeting the Sustainable Development Goals*. Rome: FAO.

Fish Farming Expert (2019) '44 Norwegian salmon farmers ditch Brazilian soy'. Available at: www.fishfarmingexpert.com/article/norwegian-salmon-farming-co-op-ditches-brazilian-soy/

Gentry, R.R., Froehlich, H.E., Grimm, D., Kareiva, P., Parke, M., Rust, M., Gaines, S. D. et al. (2017) 'Mapping the global potential for marine aquaculture', *Nature Ecology & Evolution*, vol. 1, pp. 1317–1324.

Ilaks (2019) 'Droppet brasiliansk soya: – Vi ønsker ikke å være forbundet med noe av det som skjer der borte, uansett hvor sertifisert det er'. Available at: https://ilaks.no/droppet-brasiliansk-soya-vi-onsker-ikke-a-vaere-forbundet-med-noe-av-det-som-skjer-der-borte-uansett-hvor-sertifisert-det-er/.

Kämpf, J. and Chapman, P. (2016) 'Upwelling systems of the world: A scientific journey to the most productive marine ecosystems'. Available at: https://doi.org/10.1007/978-3-319-42524-5

Khan, M.I., Shin, J.H. and Kim, J.D. (2018) 'The promising future of microalgae: Current status, challenges, and optimization of a sustainable and renewable industry for biofuels, feed, and other products'. Available at: www.ncbi.nlm.nih.gov/pmc/articles/PMC5836383/

Lindahl, O. (2011) 'Mussel farming as a tool for re-eutrophication of coastal waters: Experiences from Sweden'. Available at: www.researchgate.net/publication/230535861_Mussel_Farming_as_a_Tool_for_Re-Eutrophication_of_Coastal_Waters_Experiences_from_Sweden

Lovelock, C.E., Ball, M.C., Martin, K.C. and Feller, I.C. (2009) 'Nutrient enrichment increases mortality of mangroves'. Available at: https://journals.plos.org/plosone/article?id=10.1371/journal.pone.0005600

Lowe, A.T., Bos, J. and Ruesink, J. (2019) 'Ecosystem metabolism drives pH variability and modulates long-term ocean acidification in the Northeast Pacific coastal ocean'. Available at: www.ncbi.nlm.nih.gov/pmc/articles/PMC6353961/

MFA (Ministry of Foreign Affairs) (2017) 'The place of the oceans in Norway's foreign and development policy'. Available at: www.regjeringen.no/contentassets/1b21c0734b5042e489c24234e9927b73/en-gb/pdfs/stm201620170022000engpdfs.pdf

Mortensen, L.M. (2017) 'Remediation of nutrient-rich, brackish fjord water through production of protein-rich kelp *S. latissima* and *L. digitata*', *Journal of Applied Phycology*, vol. 29, pp. 3089–3096.

Nagothu, U.S. (2005) 'Integrated coastal zone management in Vietnam: Present potentials and future challenges', *Ocean & Coastal Management*, vol. 48, nos 9–10, pp. 813–827.

Nagothu, U.S. and Ortiz, I. (2007) 'Aquaculture in the Philippines: Socio-economics, poverty and gender'. Available at: https://pdfs.semanticscholar.org/4f87/317a9ff60bc0fb210e27ea447416757d0063.pdf

Radulovich, R., Neori, A., Valderraman, D., Reddy, C.R.K., Cronin, H. and Forster, J. (2015) 'Farming of seaweeds'. Available at: www.sciencedirect.com/science/article/pii/B9780124186972000039

Riemann, B., Carstensen, J., and Dahl, K. (2016) 'Recovery of Danish coastal ecosystems after reductions in nutrient loading: A holistic ecosystem approach', *Estuaries and Coasts* vol. 39, pp. 82–97.

SAPEA (Science Advice for Policy by European Academies) (2017) 'Food from the oceans: How can more food and biomass be obtained from the oceans in a way that does not deprive future generations of their benefits?' Berlin: SAPEA. doi:10.26356/foodfromtheoceans.

Sharma, S., Line, D., HansenJon, Ø., HansenLiv, T., Mydland Svein, J., Horn, M., Øverland, V.G.H. et al. (2018) 'Microbial protein produced from brown seaweed and spruce wood as a feed ingredient'. Available at: https://doi.org/10.1021/acs.jafc.8b01835

SINTEF (2019) Blog. Available at: https://blogg.sintef.no/sintefocean-nb/taredyrking-kan-binde-co2-i-stort-volum/

Tocher, D.R., Betancor, M.B., Sprague, M., Olsen, R.E. and Napier, J.A. (2019) 'Omega-3 long-chain polyunsaturated fatty acids, EPA and DHA: bridging the gap between supply and demand', *Nutrients*, vol. 11, no. 1, p. 89.

Tullis, P. (2017) 'How ocean aquaculture could feed the entire world – and save wild fish'. Available at: www.newsdeeply.com/oceans/articles/2017/08/16/how-ocean-aquaculture-could-feed-the-entire-world-and-save-wild-fish

UNEP (United Nations Environment Programme) (2001) 'Protecting the oceans from land-based activities'. Available at: www.jodc.go.jp/info/ioc_doc/GESAMP/report71.pdf

Ytrestøyl, T., Aas, T.S. and Åsgård, T. (2015) 'Utilisation of feed resources in production of Atlantic salmon (*Salmo salar*) in Norway', *Aquaculture*, vol. 448, pp. 365–374.

3 Land and ocean connection through iron transport by rivers

The case of the Amur-Okhotsk ecosystem (Giant Fish-Breeding Forest)

Takeo Onishi and Takayuki Shiraiwa

Introduction

The main objective of bioeconomy is to promote economic development and job creation through sustainable utilization of bio-resources. The majority of the scientific community agrees that maintenance of ecosystem services is essential for sustainable development, which is central to bioeconomy development. Ecosystem services have been categorized into provisional, supporting, regulating, and cultural services (Millennium Ecosystem Assessment, 2005). Under the category of supporting services, nutrient cycling plays an important role in the ecosystem. An adequate and balanced supply of nutrients through complex biogeochemical nutrient cycling underpins all other ecosystem services. Thus, nutrient cycling is considered the foundation of all ecosystem services.

The two key factors that function as drivers of nutrient circulation are the complex biogeochemical processes occurring all over the surface zone of the Earth, and the movement of fluids (mostly air and water), which connect the biogeochemical processes occurring in one place to another. Similar to teleconnections like El Niño and La Niña, which were found in atmospheric science, there are unknown land and ocean connections through the movement of water and nutrients. In fact, clarifying the mechanism of connections between land and oceans has been one of the most popular issues in the research field of material circulation on the Earth recently.

Since nitrogen and phosphorus are the two major nutrients and the cause of serious environmental problems, such as eutrophication, several studies have been conducted on these elements. However, not only nitrogen and phosphorus circulation, but also other nutrients, especially, the essential elements for life, are also important, and interact with each other in their circulation. Among such essential elements is iron (Fe) transport through rivers, which is one of the important systems connecting the land and the ocean ecosystem that will be the focus of this chapter. The case study introduced in this chapter might provide an interesting example showing the intimate connectivity between land and oceans.

In this chapter, a general explanation of the biogeochemical properties of iron will be provided, followed by a detailed presentation of the case of iron production and transport in the Amur-Okhotsk ecosystem. The chapter in general focuses on biogeochemical processes. In the last section of this chapter the social and economic impacts of conserving the Amur-Okhotsk system will be discussed. The chapter will highlight the concept of '*Uotsukirin*' (the fish-breeding forest) and its extended use, and the establishment and activities of the Amur-Okhotsk consortium. Finally, in the conclusion, several implications for the bioeconomy of the biogeochemical processes of the Amur-Okhotsk ecosystem will be presented.

Importance of iron for biogeochemical processes

Iron, one of the essential elements, constitutes approximately 30 per cent of the weight of the Earth. And iron is one of the most important metals used and necessary for living organisms. Moreover, in both plants and animals, iron plays an important role in their metabolic systems. In the electron transfer system within the photosynthesis processes, iron is critical for functioning in the cytochrome. In the haemoglobin and other enzymes of animals, iron also plays an important role. Most of the iron on the Earth exists in the central core of the Earth. On the other hand, on the surface of the Earth, though the iron is a very common element, it is unevenly distributed. While the amount of iron in the soil is approximately 20–40 g/kg, which is the fourth most abundant element in the soil (Cornell and Schwertmann, 2003), the average concentration of iron in freshwater rivers is estimated to be about 0.01 mg/L (Maidment, 1993). Moreover, the average concentration of iron in the sea is about 0.01 μg/kg which is 1000 times smaller than the iron concentration in rivers (Johnson, Gordon and Coale, 1997). Especially in the outer ocean, it is well known that there exists a High Nutrient Low Chlorophyll (HNLC) ocean area, where a sufficient supply of iron will stimulate phytoplankton to bloom (Boyd et al., 2000; Martin and Fitzwater, 1988; Martin et al., 1994). In other words, iron is a limiting factor of phytoplankton growth in an HNLC area. Thus, how iron is supplied to the HNLC ocean area has been an important research topic in ocean biogeochemistry. The main source of iron in the sea is from terrestrial sources. This means that through atmospheric movement or river flows, iron is slowly but gradually transported from the land to the ocean. Thus, understanding the mechanism of iron transportation from the land to the ocean is an important task for researchers engaged in studying biogeochemical processes on land.

Connection between lands and oceans

Most scientists in general agree that human activities such as agriculture, forestry and fisheries largely influence iron transportation from the land to the ocean. The supply of iron, for example, from the land due to agricultural activities might have a large effect on the amount of fish catches due to changes in the phytoplankton growth. However, more studies may be necessary to fully confirm the impacts. Thus, iron-related biogeochemical processes directly affect the fisheries and the livelihoods of millions of people across the globe. This clearly shows the need to manage ecosystems, especially the land-sea or land-ocean interactions sustainably, in order to support the biogeochemical processes and nutrient flows to promote new paradigms such as bioeconomy development. The biogeochemical cycles will not only help in promoting the economy in a given region but also have the potential to fix carbon, which is essential to reduce global warming.

Interaction between iron and soil organic matter

Since iron is abundant in the environment, and its possession of electrons in *d* orbitals with π-character (i.e. iron is active in chemical reactions with other elements) (Pauling, 1988), iron can actively react with various essential elements, such as carbon (C), nitrogen (N), oxygen (O), and sulphur (S). Among these elements, the behaviour of organic carbon is the key to understanding the element circulation. As is well recognized, soil provides a sink for a large amount of organic carbon. The total amount of soil organic carbon is estimated to be about 1,550 Gt C which is nearly three times larger than that of plant biomass, and about twice that of atmospheric carbon (Lal, 2004). This means that carbon stocked in the soil, if released into the atmosphere, is a potential source of CO_2, which is responsible for global warming. Understanding how the organic carbon is stored and how to maintain or increase carbon stocks in soils are critical for reducing global warming. Through chemical interactions between iron and soil organic carbon, iron plays a crucial role in retaining organic carbon in soils (Wagai and Meyer, 2007). At the same time, iron is mobilized by organic carbon by forming a chelating complex with organic carbon. Overall, while iron is functioning as an element to stabilize organic carbon in soils, iron is also mobilized by organic carbon at the same time. Iron thus released is leached into the oceans from the land. In addition, the microbial reduction of iron is reported to release dissolved organic carbon (Pan et al., 2016; Zhao et al., 2017). It can be said that interactions between iron and soil organic carbon regulate iron transportation. In short, understanding the behaviour of iron also contributes to a better understanding of the behaviour of carbon. Basic scientific understanding of carbon-iron interactions might lead us to find new solutions to regulate the amount of carbon stock in the soils in the future. This has great potential to contribute to carbon sequestration strategies against global warming by providing sustainable options to trap atmospheric carbon in soil.

Geographic settings of the Amur-Okhotsk system

The Amur River basin

The basin area of the Amur River is 2.1 million km^2 (Figure 3.1). The total length of the river is 4,400 km, thus making it the ninth-longest river in the world. The river basin area stretches over four countries, i.e. Mongolia, Russia, China, and North Korea. In the middle section of the river, the main course of the river forms the international border between Russia and China. The annual average discharge of the river is 8,321 m^3/s. According to the Köppen climate classification, most of the area is classified as continental climate influenced by monsoon (*Dw*). Average annual precipitation of the basin during the period from 1981 to 1990 was estimated at 490 mm (Asian Precipitation–Highly-Resolved Observational Data Integration Towards Evaluation of Water Resources: APHRODITE Water Resources, Yatagai et al., 2009), ranging from 200 mm in the upstream to 700 mm in the middle and downstream (Simonov and Dahmer, 2008). Since the area stretches over 15 ° from south to north, the annual average temperature ranges from -7 °C to 6 °C. Due to the low temperatures in winter season, significant sections of the Amur River are frozen during the period from November to April.

The land-use/land cover type of the basin is comprised of forest (59.5 per cent), agricultural land (18.3 per cent), grassland (12.2 per cent), and wetland (6.9 per cent) (Yermoshin et al., 2007). The main forest types are mixed forest and coniferous forest, and the dominant tree species are Dahurian Larch (*Larix gmelinii*), Ezo spruce (*Picea jezoensis*) and Siberian silver birch (*Betula platyphylla*). During the past 100 years, the land-use/land cover type has drastically changed. In particular, the wetlands have been converted to agricultural lands in China. When we compare land-use/land cover types between the 1930s and 2000, the wetland area has decreased from 13.3 per cent to 6.9 per cent. This decrease was mainly attributed to the conversion of wetlands to agricultural land (especially to upland, not to paddy fields) in China, because of the rapid population increase and economic growth in China after the Second World War. In addition, the rapid economic growth in China is also accelerating the import of huge numbers of logged trees from the Russian territory, though the total forest area as shown in the statistical data has not significantly changed between the 1930s and 2000. Furthermore, forest fires happening in the Amur River are attributed to logging, and thus the impact of logging on the ecosystem of this region is another concern.

The Sea of Okhotsk and the Oyashio Region

Freshwater discharged from the Amur River flows into the Sea of Okhotsk through the Sakhalin Bay (see Figure 3.1). Though the river

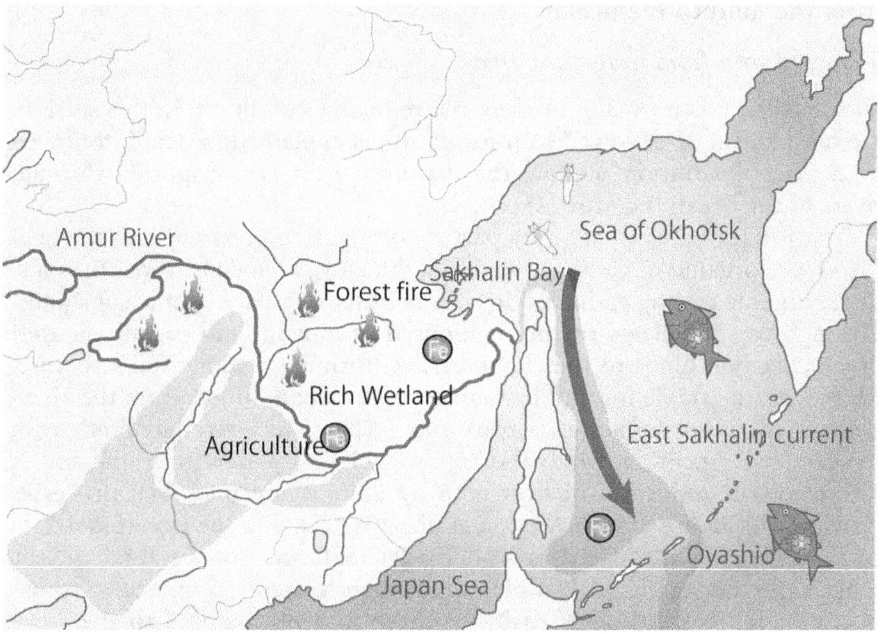

Figure 3.1 Geographic setting of the Amur and Okhotsk ecosystem.
Source: modified after Kawaguchi (2007).

mouth is also connected to the Sea of Japan through the Strait of Tartary, geographic conditions make the freshwater flow in a northernly direction. The area of the Sea of Okhotsk is about 1.6 million km^2, comparably as large as the Amur River basin area. The average depth of the sea is 859 m, and the deepest depth is 3,372 m. The Sea of Okhotsk is a marginal sea of the Western Pacific Ocean. The sea is bounded by the Kamchatka Peninsula on the east, Kuril Island on the south-east, Hokkaido Island on the south, and the Sakhalin along the west. Since the cold pole in the Northern Hemisphere is located upwind of the Sea of Okhotsk, the sea is the southern limit of the sea ice in the Northern Hemisphere. At the southern boundary, the Sea of Okhotsk is connected to the Oyashio region through the Kuril Islands. The Sea of Okhotsk and the Oyashio region are one of the fields where fish productivity is high. Since this region is one of the HNLC ocean areas, iron supply is essential to support the high biological productivity of this region. Due to the recent climate change and natural variability of the climatic conditions, the dominant fish species caught for sale are changing periodically. Thus, establishment of fisheries which can flexibly adapt to the ocean environment is the key challenge in this region (Sakurai, 2007).

Mechanisms of dissolved iron generation and transportation from the land to the ocean

Source of iron from terrestrial areas

Dissolved iron is normally formed in a reductive condition. In this section, the mechanism of dissolved iron formation is explained in detail, followed by a brief illustration of how the dissolved iron is transported through rivers to the ocean (Figure 3.2).

Iron usually consists of a clay particle originating in parent rocks, which forms an organic-mineral complex, stabilizing organic carbon in soils. Here, organic carbon stabilized in soils is usually called a humic substance. Through the complex structure composed of mineral and organic matter, organic carbon binds to the clay particles, forming a stable soil structure. Moreover, the stable form of iron under natural conditions, from the viewpoint of thermodynamics, is ferrous iron Fe(Ⅲ), which strongly binds with oxygen and precipitates into the soil particle. However, once the soil is saturated, the reduction of iron both by abiotic and microbial processes (Lovley and Philips, 1988; Melton et al., 2014) makes the iron dissolve in water (changing from ferrous iron Fe(Ⅲ) to ferric iron Fe(Ⅱ)). On the other hand, under the normal pH and oxygen concentration usually found in terrestrial water, iron is easily oxidized and precipitated to the soils.

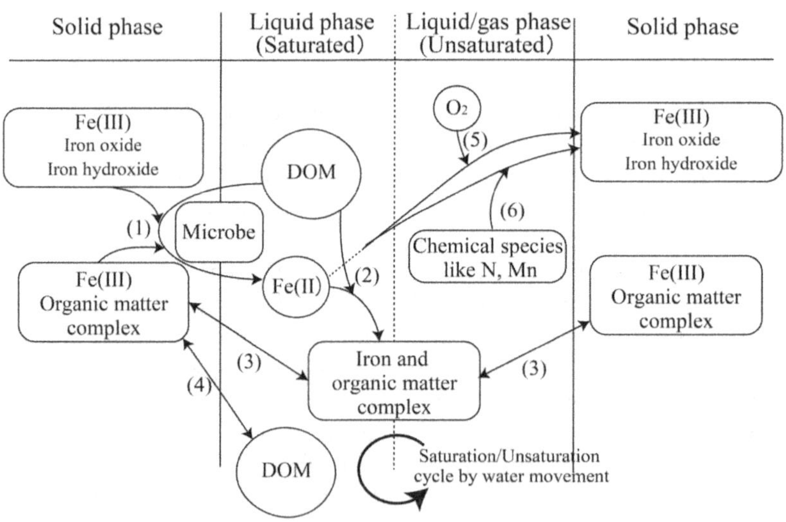

(1): Iron reduction by microbes, (2): Chelation between dissolved organic matter (DOM) and Fe(II),
(3): Adsoption/Desorption reaction between soil particle and iron organic matter complex,
(4): Adsoption/Desorption reaction between soil particle and DOM
(5): Oxidization of Fe(II) by oxygen and adsorption reaction to soil particle,
(6): Oxidization of Fe(II) by N, Mn chemical species and adsorption reaction to soil particle

Figure 3.2 Biogeochemical processes related to iron behaviour in soils.

Thus, dissolved iron cannot exist in water for a long time, even if the iron dissolves in the water. However, if the iron is chelated with organic matter mainly by a coordinate bond, the chelated form of iron can be stably dissolved in water. Thus, in short, iron reduction and chelation with organic matter are the two fundamental processes to produce the dissolved form of iron in soils. Important biogeochemical processes related to iron behaviour in soils are schematically explained in Figure 3.2.

Considering the above-mentioned biogeochemical mechanism working in soils, we can expect that wetlands will have a high potential to form dissolved iron. Since wetlands have abundant water, consequently, they are rich in organic carbon typically found as a form of peat, which is rich in humic substances. As is well known, dissolved iron concentration observed in wetlands is significantly higher than other land-use types such as paddy fields, uplands, and forests. As Yoh et al. (2010) revealed, dissolved iron concentration observed in wetlands is about 10 times higher than that of forests. Here, it should be noted that forest area occupies more than half of the total area of the Amur River basin. Thus, the total amount of iron flowing out from forests is as important as that from wetlands. Since the dissolved iron concentration observed in uplands is very low, land-use conversion from wetlands to agricultural lands is expected to have a great impact on the dissolved iron flux. On the other hand, the dissolved iron concentration in paddy fields is much higher than that in uplands, but slightly lower than that in wetlands. To estimate how land-use conversion will change the flux of dissolved iron, a hydro-chemical model can help us to illustrate the impacts. The results of the model evaluation will be explained in the next section.

Modelling dissolved iron transport in the Amur River basin

The structure of the hydro-chemical model developed by Onishi et al. (2010) is illustrated in Figure 3.3. The model is composed of two modules. One is for hydrological simulation and the other is for the simulation of dissolved iron formation in the soils. The basic structure of the hydrological model is based on the TOPMODEL concept (Beven and Kirkby, 1979). The model structure of the dissolved iron formation was developed by Onishi et al. (2010). The whole basin was sub-divided into 1,036 grids. The area of each grid is approximately 2,000–2,500 km^2. Each sub-divided grid was considered to be a sub-watershed, from which discharge is routed to one of eight surrounding grids (sub-watersheds) according to the flow direction assigned to the river flow network of the Amur River basin. Since the Amur River basin is relatively flat, in our calculations, river flow velocity used for flow routing was set as constant at 0.5 m/s. Next, each sub-watershed was again sub-divided into a finer grid with a grid size of 1 km^2. In each grid, all the essential hydrological processes shown in Figure 3.3 were calculated. Through this process, the amount of soil water content of

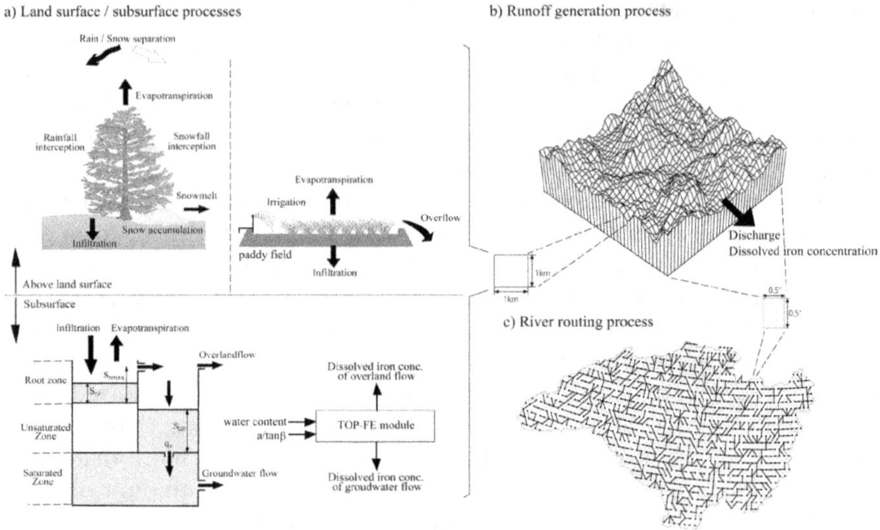

Figure 3.3 Schematic explanation of the hydro-chemical model.
Source: Onishi et al. (2010).

each grid at any time can be calculated. This value can then be used to calculate the dissolved iron concentration at that time.

Here, another index called the topographic index of each grid point was introduced to calculate the dissolved iron concentration. The topographic index is defined as $a/\tan\beta$, here, a is the watershed area of a point, and $\tan\beta$ is a slope. The index indicates the wetness (or dryness) of each grid point. Thus, a land cover type like wetland tends to have a higher value than other land cover types. Onishi et al. (2010) found that the average dissolved iron concentration of a watershed can be correlated with the topographic index based on a statistical analysis of 43 different watersheds. Based on this relationship between the dissolved iron concentration and the topographic index, the dissolved iron concentration of each grid was calculated. After the calibration and validation procedures, we applied the model to evaluate the effect of the land cover conversion on the dissolved iron flux of the basin. The results are shown in Figure 3.4. It indicates that if the wetland is completely converted to agricultural lands, the dissolved iron flux of the basin is decreased by about 30 per cent compared to the present condition. However, there remains much room for improvement in the accuracy of the estimated values, but the estimates do suggest the importance of conserving wetlands. However, how the change in dissolved iron flux will impact phytoplankton growth in the Sea of Okhotsk cannot be fully explained by this study.

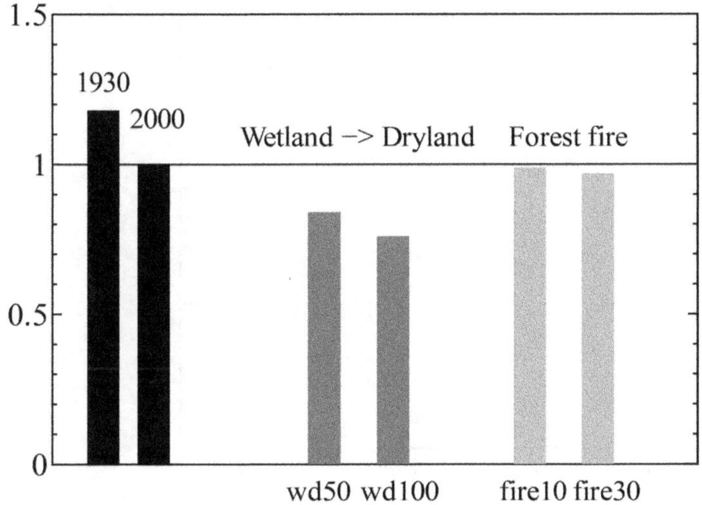

Figure 3.4 Simulated results of change in dissolved iron flux by land cover conversion.

Source: Onishi and Yoh (2009).

Notes

The result was standardized against the dissolved iron flux under the land cover condition prevalent in 2000. Bogorodoskoye is the name of the city located near the mouth of the Amur River.
1930: landcover condition in 1930;
2000: landcover condition in 2000;
wd50: 50% of wetland was converted to dryland;
wd100: 100% of wetland was converted to dryland;
fire10: 10% of forest was affected by forest fires;
fire30: 30% of forest was affected by forest fire.

Long-term trend of dissolved iron concentration in the Amur River basin

While our model can calculate the average behaviour of dissolved iron flux in the basin, dissolved iron concentration seems to show a drastic multi-decadal change during the past 40–50 years. As an example, Figure 3.5 shows the dissolved iron concentration observed at the Khabarovsk observation station. During the late 1990s, a large increase in dissolved iron concentration was recorded, and the same trends can be confirmed throughout the basin.

To clarify the mechanism behind this increase in dissolved iron concentration, we hypothesized that the effect of the soil freezing and thawing processes on dissolved iron generation might play an important role (Shamov et al., 2014). Our three years' field study during the period from

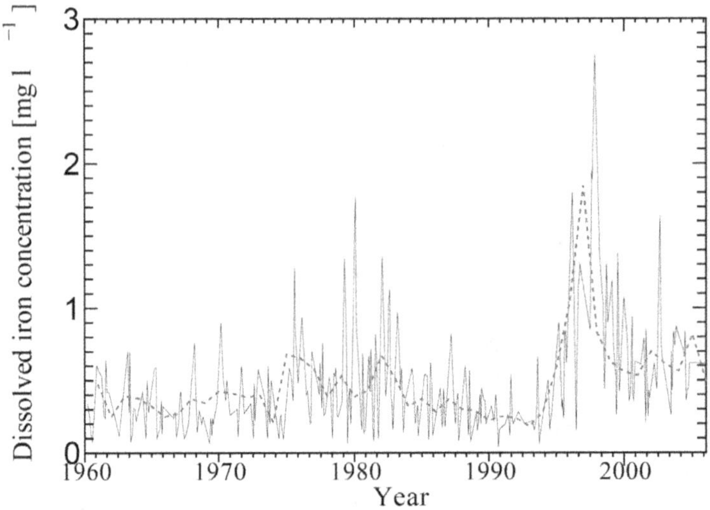

Figure 3.5 Long-term dissolved iron concentration at the Khabarovsk observation station.

Source: adapted from Shamov et al. (2014).

Note

The dashed line is the 5-year moving average of the original monthly data.

2015 to 2017 in the Tyrma River basin, one of the largest tributaries of the Bureya River flowing into the main course of the Amur River, has revealed the following facts:

1 In the riparian zone especially along meandering rivers, a unique landscape called 'марь' (in Russian, pronounced in English as Mar) is dominant.
2 Марь is characterized by vegetation typically found in wetlands and sparsely populated Chinese Pine.
3 The main soil type is peat, and we could always find permafrost in the peat layer and the underlying mineral soil layer.
4 Affected by the depth of the active layer of the permafrost, dissolved iron concentration of the марь changed significantly.
5 Especially when the thawing depth of permafrost reaches the underlying mineral layer, the dissolved iron concentration of the soil tends to increase.
6 Long-term analysis of thawing depth of the study site using climate data and the Stefan solution which can give a good estimate of thawing depth showed that the maximum thawing depth and thawing duration at the mineral soil were highest during the late 1990s.

Taking these findings into consideration, we can conclude that thawing of the permafrost in марь could have contributed to the increase in the dissolved iron concentration and the flux of rivers. On the other hand, upscaling the results obtained in a small catchment to the whole basin scale is a bit premature. Upscaling procedures are very challenging and always pose questions for hydrological researchers. However, we expect that with the aid of remotely sensed data, and hydro-chemical modelling, we can extend the measured data collected from small-scale catchments to much larger areas. If we are successful in upscaling, we could provide a better explanation for the question: how does climate change affect the flux of dissolved iron in the Amur River basin? Thus, further research is required to prove the effect of freezing and thawing of permafrost on the dissolved iron behaviour in soils at the basin scale.

Transportation process in rivers, estuaries, and oceans

In principle, during the transportation by river waters, the biogeochemical processes will affect the form of the dissolved iron in river waters. Thus, we have to consider the transformation of the dissolved iron in river systems. The possible mechanisms of the transformation of the dissolved iron are coagulation and deposition on the riverbed, the dissolution of dissolved iron from the riverbed or lake bottom sediments, light decomposition of organic matter, and utilization by phytoplankton in rivers or lakes. So far, only a few studies have been conducted on dissolved iron behaviour in rivers. Nagao et al. (2010) conducted a survey along the different points of the middle stream and downstream of the main course of the Amur River. Their study showed that there was no significant change in dissolved iron concentration along the river. Considering that there was no significant inflow from the tributaries along the transect line of the survey, it can be concluded that the net amount of dissolved iron in the Amur River has not significantly changed during transportation.

In the estuary zone, how the dissolved iron is affected by the mixing of freshwater and seawater has been surveyed for a long time (Boyle et al., 1977). In the case of the Amur River, Terashima and Nagao (2007) found that almost 90 per cent of dissolved iron was removed from the river water by coagulation and deposited as sediment in the estuary zone of the river. On the other hand, this removal rate was not constant and varied from 50–90 per cent. This can be attributed to the fact that sea water with high salinity strongly influences coagulation and thereby increases the dissolved iron sediment material. The characteristics of organic matter that influence chelating with iron are also important in deciding the removal rate. Once the dissolved iron is removed from water, it is deposited in the coastal areas. Thus, commonly, most of the dissolved iron transported from rivers will not be further transported into seas. However, in the case of the Sea of

Okhotsk, a unique transporting mechanism exists characterized by a combination of the thermohaline circulation and tidal mixing.

As addressed in the previous section, the Sea of Okhotsk is covered by thick sea ice during the winter seasons, since the area is near to the cold pole of the Northern Hemisphere. As a result of the cold wind blowing from the Siberian high-pressure area, the sea becomes frozen. In addition to this, huge amounts of freshwater delivered from the Amur River also play an important role in the formation of sea ice. Consequently, during the process of freezing, dense cold water is formed under the sea ice. Since this mass of dense water is heavier than the surrounding sea water, it goes down to the deeper part of the sea and generates a large-scale convective motion in the sea. The ocean current formed in the Sea of Okhotsk is called the North Pacific Intermediate Water (NPIW) (Nishioka et al., 2013; Shiraiwa, 2012). Generally, ocean circulation induced by this process is called thermohaline circulation. The typical case can be found in the northern Atlantic Ocean, which is considered the starting point of the global oceanic circulation commonly called Broecker's conveyer belt. Another important physical mechanism in the Sea of Okhotsk is tidal mixing. Even if the thermohaline circulation exists, if the iron deposited on the sea floor is not incorporated into this circulation, iron will not be transported into the oceans. Thus, tidal mixing also plays an important role. Since the range of tidal fluctuation mainly decides the magnitude of this tidal mixing effect, the Sea of Okhotsk, where the tidal fluctuation range is large, has a large tidal mixing. As a result, sediments deposited on the sea floor rise up through this tidal mixing and are incorporated into the thermohaline circulation.

Source of iron to oceans: debate on sediments or air dusts

Historically, air dust has been considered the main source of iron provided to oceans (Martin and Fitzwater, 1988). Particularly in the outer oceans, ocean currents are not necessarily transporting dissolved iron from coastal zones to outer oceans. However, it is not yet certain how important atmospherically derived iron is to the primary production of oceans mainly because of its insoluble nature (Nishioka et al., 2013). On the other hand, a recent study using numerical simulations shows that iron deposited on the continental shelf that originated from the land and iron transported by air dusts are both considered to be equally important sources of iron in oceans (Moore and Braucher, 2008). Active discussions on the source of dissolved iron in oceans is ongoing. In the case of the Sea of Okhotsk, dissolved iron from sediments and air dust are both considered important. According to Nishioka et al. (2013), while dissolved iron transported by the ocean current called the NPIW is supporting the stable supply of dissolved iron to the Sea of Okhotsk and especially to the Oyashio region, iron transported by air dust will give a sporadic supply of a significant amount of dissolved iron to the Oyashio region.

Summary of iron circulation in the Amur-Okhotsk system

The characteristics of iron circulation in the Amur-Okhotsk system are summarized below and the schematic illustration is given in Figure 3.6:

1 The most important mechanism of dissolved iron generation in the soils is the formation of a reduced condition and chelating with organic matter, such as humic substances. Thus, a wetland-type vegetation area has the highest potential to generate dissolved iron. On the other hand, even if the dissolved iron concentration is low, forest-type vegetation also contributes to the total dissolved iron flux because its area occupies 60 per cent of the total area of the Amur River basin.
2 Most (approximately 90 per cent) of the dissolved iron transported through the Amur River is removed by the coagulation process at the estuary of the Amur River. However, due to the combined effects of the ocean current called the NPIW, which is created by sea ice formation and tidal mixing, iron deposited on the continental sea floor, including the Gulf of Sakhalin, is again transported to the Oyashio region, supplying huge amounts of iron to this region.
3 Based on the numerical simulations of the hydro-chemical model we constructed, wetland conversion to agricultural lands might significantly decrease the amount of dissolved iron produced in the Amur River basin. This in turn might significantly influence the fisheries sector in the region demonstrating the implications of landscape changes on the biogeochemical processes.

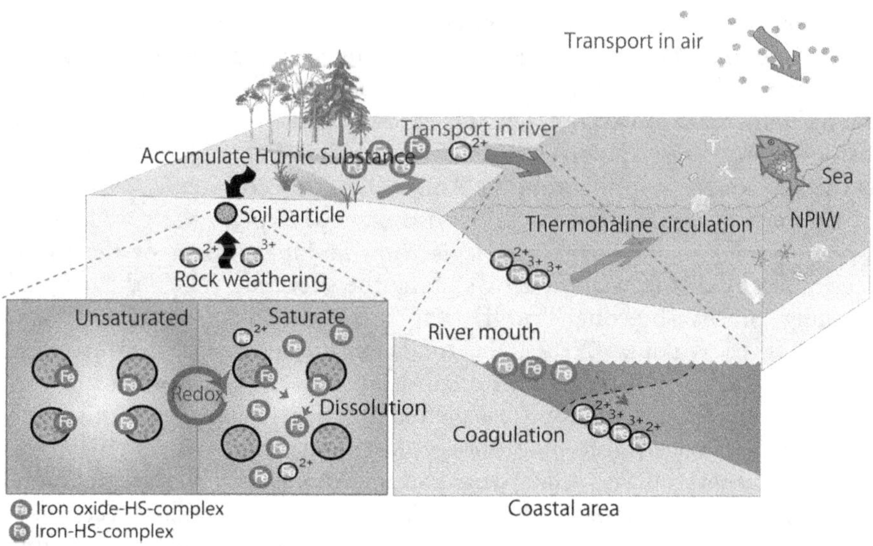

Figure 3.6 Schematic illustration of dissolved iron circulation in the Amur–Okhotsk ecosystem.

Initiatives to conserve the Amur-Okhotsk ecosystem

Concept of fish-breeding forest

Considering the above-mentioned mechanisms of dissolved iron formation and transportation from land to ocean through rivers, we can conclude that the Amur River and the Sea of Okhotsk and the Oyashio region constitute one large ecosystem encompassing lands and oceans. This can be termed '*Uotsukirin*' (in Japanese), meaning 'fish-breeding forest', which very well explains the functionality of the system (the term 'fish-breeding forest' will be used from now on).

However, if we look at the history of the 'fish-breeding forest' in Japan, we find that there is also a distinct difference between the original meaning and our use of 'fish-breeding forest' for the Amur-Okhotsk ecosystem. The concept of a 'fish-breeding forest' originally emerged from Japanese indigenous knowledge that can be dated back to the early stages of the Edo Period (1603–1868) (Wakana, 2012). Originally, the term 'forest' in the expression specifically indicated trees growing along the shoreline. Although the underlying mechanisms were not clearly understood at that time, people in the Edo period empirically found that fish shoals tend to gather at the shoreline where trees grow. On the other hand, in the case of the Amur-Okhotsk ecosystem, the mechanism which connects the forest (but more accurately, the wetlands) and the ocean was scientifically studied, and a significant part of the mechanism was clarified scientifically. Moreover, 'forest' in the term means not only trees along shorelines, but also indicates all forests and wetlands existing in the basin. Thus, literally speaking, we modified the original meaning indicated by 'forest' in *Uotsukirin* to suit the Amur-Okhotsk ecosystem. In addition, in the case of the Amur-Okhotsk system, the system extends over a very large area, and the mechanism is more complex. Thus, we added the word 'giant' before 'fish-breeding forest'. Finally, 'Giant Fish-Breeding Forest' (GFBF) emerged as the key concept representing the Amur-Okhotsk ecosystem.

The term 'fish-breeding forest' is also used to express the specific conservation zone in modern forest legislation in Japan. Most of the forest designated by this legislation grows along the shoreline. Thus, the original meaning of 'fish-breeding forest' was also inherited by modern law. According to Wakana (2012), the total area designated as fish-breeding forest in Japan is about 580 km^2, comprising only 0.5 per cent of the protected forest. Though the total designated area is not large, how this legislation can influence future bioeconomy development-related activities such as coastal fisheries is very interesting and important. Tracking the relationships between traditional indigenous knowledge and modern economic activities might give a useful insight into the future bioeconomy planning and development along the coastal areas. In the process, it may influence the communities to consider environmental protection as one of their priorities. From the viewpoint of environmental philosophy, various concepts and ideas related to the connection between lands and rivers, and lands and oceans emerged during the Edo Period. This could be an interesting topic in itself for researchers in the field of environmental philosophy. The concept was actually revived by a Japanese fisherman (also an environmental activist) in the 1980s, using the slogan 'the forest is the sweetheart of the sea' (Hatakeyama, 2006). Coincidently, the research community has also started to pay attention to this concept in recent years and since then several studies have been conducted to explain the connection between land, fisheries and oceans (Shimizu et al., 2014). Among these research initiatives, establishing connectivity within and between ecosystems and humans (CoHHO, www.cohho.kyoto-u.ac.jp/en/), led by Kyoto University, is an ongoing comprehensive research and educational programme that is worth mentioning here.

Establishment of the Amur-Okhotsk consortium

Based on several decades of accumulating research, conducted by researchers from various disciplines from the natural to the social sciences, awareness of the importance of the GFBF ecosystem has gradually increased. Among the several research projects to clarify the mechanism of the GFBF ecosystem, the Amur-Okhotsk project, an international project involving researchers from China, Russia, and Japan, funded by the Research Institute for Humanity and Nature, conducted from 2006 to 2010, has produced interesting results. As part of the activities of the Amur-Okhotsk project, the Amur-Okhotsk Consortium was established in November 2009. The consortium has organized bi-annual meetings for the last 10 years on a regular basis. In the following section, the objective and the reasons for establishment of the consortium are explained.

In the region where the Amur-Okhotsk ecosystem lies, there has been serious political tension between the countries sharing the ecosystem. Though some bilateral environmental agreements between several countries exist, there has been no multilateral cooperative framework under which relevant countries could discuss the challenges and conservation initiatives of the GFBF ecosystem. To foster a multilateral international cooperative framework, the Amur-Okhotsk Consortium was established as a first step. The purpose of the consortium was to provide an international platform to network and discuss the conservation issues and the sustainable use of the GFBF ecosystem. The consortium members comprised research institutes, universities, government organizations, international organizations and private companies (Hanamatsu, 2012). In the past four meetings, various issues were discussed, including procedures to establish and develop sustainable ways of maintaining consortium cooperation. At this moment, the consortium is mainly focused on academic and research activities. The challenge is to keep the consortium active and to include a much wider range of stakeholders in the future. Such platforms help to provide a common framework to manage sensitive ecosystems sustainably as we promote bioeconomy in the regions.

Conclusion

The Amur-Okhotsk ecosystem, the so-called Giant Fish-Breeding Forest, is a complex and unique natural system that provides not only ecosystem services but also supports fisheries in the region. The following is a summary of the main findings of the scientific research undertaken during the last decade on the Amur-Okhotsk ecosystem:

1 In the Amur River basin, wetland-type vegetation area has the highest potential to generate dissolved iron. Although most of the dissolved iron transported through the Amur River is removed at the estuary of the Amur River, due to the combined effects of the ocean current called the NPIW, which is formed by sea ice formation and tidal mixing, the iron deposited on the continental sea floor, including the Gulf of Sakhalin, is again transported back to the Oyashio region, supplying huge amounts of iron to this region.
2 Based on the numerical simulations of the hydro-chemical model we constructed, wetland conversion to agricultural lands might significantly decrease the amount of dissolved iron produced in the Amur River basin. According to several different scientific investigations, it is highly possible that significant amounts of dissolved iron are transported from the land to the ocean and are supplied in the Oyashio region. However, we do not have direct evidence yet how the changes in land conditions will affect the productivity of phytoplankton in the Oyashio region. This also might influence carbon sequestration in the long run, showing the influence of improper land-use planning on carbon sinks and global warming.

Thus, more research is needed to establish further evidence to support the connectivity between land and oceans in other regions worldwide. At this moment, we are able to study only the impacts due to rapid changes such as land cover changes and forest fires in the basin. In addition to these changes, climate change impacts on the production and transport of dissolved iron are not fully understood. According to our study, the effect of freezing and thawing on the production of dissolved iron, and temperature rise as a result of climate change, might increase the rate of dissolved iron production from the Amur River basin. Consequently, the amount of iron transported to the ocean is expected to increase. This, in turn, might foster the growth of phytoplankton in the Oyashio region. However, it is not as easy as simply saying that this has only positive effects.

The reason behind the above-mentioned statement is that there is too much uncertainty about the change of phytoplankton growth induced by the change of iron supply. An experiment using an artificial iron supply in the oceans was conducted in the 1990s but such experiments are now prohibited because of their uncertain outcomes (Buesseler et al., 2008). The implications for bioeconomy development in this region due to excess supply of iron to oceans are not clear due to uncertainty. In other words, whenever we want to foster phytoplankton growth by enhancing the iron supply, which will serve to enhance fisheries, we should be very careful about the impacts on the stability of the ecosystem.

Another important point for the bioeconomy in the Amur-Okhotsk ecosystem is that there exists a trade-off between agriculture and fisheries. As already explained in detail, wetland conversion to agricultural land (especially uplands) will significantly decrease the dissolved iron flux. On the other hand, it should be noted that there is a significant difference in the dissolved iron production rate between the uplands and the paddy fields. Since the seasonal water regime of paddy fields is similar to wetlands, the dissolved iron production rate of paddy fields is much higher than that of the uplands, but slightly lower than that of the wetlands. This means that paddy fields can be a surrogate for wetlands in terms of dissolved iron production. Thus, conversion of uplands to paddy fields is argued by some as one possible way to mitigate the trade-off between agricultural activities and fisheries through maintaining the dissolved iron production rate from the terrestrial area.

In the case of the Amur River, there is a possibility that excessive agricultural activities mainly happening on the Chinese side will have a negative influence on the fisheries of the Oyashio region by decreasing the iron flux. A much more interesting point is that the problem structure affects not only the Amur-Okhotsk ecosystem, but also similar ecosystems where the land and ocean interactions occur. One of the common features we find in environmental problems happening in a watershed is an asymmetry between upstream stakeholders and downstream stakeholders. Since water is always flowing from upstream to downstream, any negative impact on a watershed is always propagated from the point where the impact occurs on the downstream. Since upstream stakeholders are less influenced by the downstream negative impacts, upstream stakeholders always tend to be insensitive to the downstream conditions. This is also the case for the Amur-Okhotsk ecosystem.

What makes the Amur-Okhotsk ecosystem different from other similar land and ocean interaction system is its spatial scale. In the case of the Amur-Okhotsk ecosystem, it extends over more than 3.6 million km^2, covering four different countries. Since the spatial scale of the system is so large, it is difficult to manage conflicts among many different stakeholders. Moreover, how this Amur-Okhotsk ecosystem responds to external forces, including human activities and global warming, is a complex issue. And it is expected that the responding time scale of the system might be sometimes a decade or several decades. In other words, we will have to wait for a long time before any external impact is reflected by the response of the system. This means the time scale is also beyond the scale of our normal economic system (at least, the modern economic system). Thus, it is a very challenging task to modify the modern economic system to adapt with the natural variability of the Amur-Okhotsk ecosystem in this region.

At this moment, the Amur-Okhotsk consortium is only serving as an academic platform to enhance and foster international cooperative research on the Amur-Okhotsk ecosystem. Though many difficult problems remain unsolved, the Amur-Okhotsk consortium might also help to enhance communication between the different stakeholders from the four countries and generate new knowledge that can contribute to a better understanding of the land-ocean interactions. Based on the interests and mutual understanding of each stakeholder (especially stakeholders related to the agriculture and fisheries sectors), construction of a common recognition of the Amur-Okhotsk ecosystem is the only way to achieve sustainable use of the bioresources of this region.

Acknowledgements

The major part of this work was conducted under the framework of an international project entitled 'Human activities in Northeastern Asia and their impact on the biological productivity in North Pacific Ocean', funded by the Research Institute for Humanity and Nature (RIHN). A part of the work was also conducted through funding support from JSPS Kakenhi No. 15H05208.

The authors express their thanks to Global Soil Data Task, the IGBP, ORNL DAAC and ISLSCP Initiative II for providing soil characteristic data. We are very grateful to all the staff in ROSHYDROMET, particularly, G. Alexander for his special collaboration on this work.

References

Beven, K.J. and Kirkby, M.J. (1979) 'A physically based variable contributing area model of basin hydrology', *Hydrological Science Bulletin*, vol. 24, pp. 43–69.

Boyd, P.W., Watson, A.J., Law, C.S., Abraham, E.R., Trull, T., Murdoch, R., Bakker, D.C.E., et al. (2000) 'A mesoscale phytoplankton bloom in the polar Southern Ocean stimulated by iron fertilization', *Nature*, vol. 407, pp. 695–702.

Boyle, E.A., Edmond, J.M. and Sholkovitz, E.R. (1977) 'The mechanism of iron removal in estuaries', *Geochimica et Cosmochimica Acta*, vol. 41, pp. 1313–1324.

Buesseler, K.O., Doney, S.C., Karl, D.M., Boyd, P.W., Caldeira K., Chai, F., Coale, K.H., et al. (2008) 'Ocean iron fertilization: Moving forward in a sea of uncertainty', *Science*, vol. 319, p. 162.

Cornell, R.M. and Schwertmann, U. (2003) *The Iron Oxides: Structure, Properties, Reactions and Uses*, 2nd edn. Weinheim: Wiley VCH.

Hanamatsu, Y. (2012) 'National boundaries and the fragmentation of governance systems: Amur-Okhotsk ecosystem from the legal and political perspective'. In M. Taniguchi and T. Shiraiwa (eds) *The Dilemma of Boundaries*. Tokyo: Springer Japan, pp. 123–143.

Hatakeyama, S. (2006) *The Forest Is Sweetheart of the Sea*. Japan: Bungei Shunjyu Publisher.

Johnson, K.S., Gordon, R.M. and Coale, K.H. (1997) 'What controls dissolved iron concentrations in the world ocean?' *Marine Chemistry*, vol. 57, pp. 137–161.

Lal, R. (2004) 'Soil carbon sequestration impacts on global climate change and food security', *Science*, vol. 304, no. 5677, pp. 1623–1627.

Lovley, D.R. and Phillips, E.J.P. (1988) 'Novel mode of microbial energy metabolism: organic carbon oxidation coupled to dissimilatory reduction of iron or manganese', *Applied Environmental Microbiology*, vol. 54, pp. 1472–1480.

Maidment, D.R. (1993) *Handbook of Hydrology*. London: McGraw-Hill Education.

Martin, J.H., Coale, K.H., Johnson, K.S., Fitzwater, S.E., Gordon, R.M., Tanner, S.J., Hunter, C.N., et al. (1994) 'Testing the iron hypothesis in ecosystems of the equatorial Pacific Ocean', *Nature*, vol. 371, pp. 123–129.

Martin, J.H. and Fitzwater, S.E. (1988) 'Iron deficiency limits phytoplankton in the northeast Pacific subarctic', *Nature*, vol. 311, pp. 341–343.

Melton, E.D., Swanner, E.D., Behrens, P., Schmidt, S.C. and Kappler, A. (2014) 'The interplay of microbially mediated and abiotic reactions in the biogeochemical Fe cycle', *Nature Microbiology*, vol. 12, pp. 797–808.

Millennium Ecosystem Assessment (2005) *Ecosystems and Human Well-being: Synthesis*. Washington, DC: Island Press.

Moore, J.K. and Braucher, O. (2008) 'Sedimentary and mineral dust sources of dissolved iron to the world ocean', *Biogeosciences*, vol. 5, pp. 631–656.

Nagao, S., Terashima, M., Seki, O., Takata H., Kawahigashi, M., Kodama, H., Kim, V.I., et al. (2010) 'Biogeochemical behavior of iron in the lower Amur River and Amur-Liman', *Annual Report on Amur-Okhotsk Project*, no. 6, pp. 41–50.

Nishioka, J., Nakatsuka, T., Watanabe, Y.W., Yasuda, I., Kuma, K., Ogawa, H., Ebuchi, N., et al. (2013) 'Intensive mixing along an island chain controls oceanic biogeochemical cycles', *Global Biogeochemical Cycles*, vol, 27, no. 3, pp. 920–929.

Onishi, T. and Muneoki, Y. (2009) 'Evaluation of landcover change impacts on dissolved iron flux of the Amur River', *Geography*, vol. 54, no. 12, pp. 2–8.

Onishi, T., Muneoki, Y., Shibata, H., Nagao, S., Kawahigashi, M. and Shamov, V.V. (2010) 'Topography as a macroscopic index for the dissolved iron productivity of different land cover types in the Amur River Basin', *Hydrological Research Letters*, vol. 4, pp. 85–89.

Pan, W., Kan, J., Inamdar, S., Chen, C. and Sparks, D. (2016) 'Dissimilatory microbial iron reduction release DOC (dissolved organic carbon) from carbon-ferrihydrite association', *Soil Biology and Biochemistry*, vol. 103, pp. 232–240.

Pauling, L. (1988) *General Chemistry*. New York: Dover Publications.

Sakurai, Y. (2007) 'An overview of the Oyashio ecosystem', *Deep Sea Research Part II: Topical Studies in Oceanography*, vol. 54, pp. 2526–2542.

Shamov, V.V., Onishi, T. and Kulakov, V. V. (2014) 'Dissolved iron runoff in Amur basin rivers in the late XX century', *Water Resources*, vol. 41, pp. 201–209.

Shimizu, N., Tateno, R., Kasai, A., Mukai, H. and Yamashita, Y. (2014) *Connectivity of Hills, Humans and Oceans*. Kyoto: Kyoto University Press.

Shiraiwa, T. (2012): 'Giant Fish-Breeding Forest: A new environmental system linking continental watershed with open water'. In M. Taniguchi and T. Shiraiwa (eds), *The Dilemma of Boundaries*. Tokyo: Springer Japan, pp. 73–85.

Simonov, E.A. and Dahmer, T.D. (2008) *Amur-Heilong River Basin Reader*. Hong Kong: Ecosystems Ltd.

Terashima, M and Nagao, S. (2007) 'Removal and fractional characteristics of dissolved iron in estuarine mixing zone', *Annual Report on Amur Okhotsk Project*, no. 4, pp. 69–74.

Wagai, R. and Mayer, L.M. (2007) 'Sorptive stabilization of organic matter in soils by hydrous iron oxides', *Geochimica et Cosmochimica Acta*, vol. 71, pp. 25–35.

Wakana, H. (2012) 'History of "Uotsukirin" (fish-breeding forest) in Japan', In M. Taniguchi and T. Shiraiwa (eds), *The Dilemma of Boundaries*. Tokyo: Springer Japan, pp. 145–160.

Yatagai, A., Arakawa, O., Kamiguchi, K., Kawamoto, H., Nodzu, M.I. and Hamada, A. (2009) 'A 44-year daily gridded precipitation dataset for Asia based on a dense network of rain gauges', *SOLA*, vol. 5, pp. 137–140.

Yermoshin, V.V., Ganzei, S.S., Murzin, A.V., Mishina, N.V. and Kudryavtzeva, E.P. (2007) 'Creation of GIS for Amur River basin: The basic geographical information', *Annual Report of Amur-Okhotsk Project*, no. 4, pp. 151–159.

Yoh, M., Shibata, H., Onishi, T. Kawahigashi, M., Guo, Y., Yamagata, K., Shamov, V.V., et al. (2010) 'Iron dynamics in terrestrial ecosystems in the Amur River basin', *Annual Report on Amur-Okhotsk Project*, no. 6, pp. 51–62.

Zhao, Q., Adhikari, D., Huang, R., Patel, A., Wang, X., Tang, Y., Obrist, D., et al. (2017) 'Coupled dynamics of iron and iron-bound organic carbon in forest soils during anaerobic reduction', *Chemical Geology*, vol. 464, pp. 118–126.

4 Evaluating local sustainability, including ecosystem services provided by rural areas to cities to promote bioeconomy

Tohru Nakashizuka, Kunihiko Kobayashi, Rei Shibata, Masahiro Aiba, Takahiro Sasai Michio Oguro, Hiroko Kurokawa and Shunsuke Managi

Introduction

One of the important aspects of the bioeconomy is to attain sustainability at various spatial scales through the judicious use of renewable resources. The concept of sustainability covers various aspects, as seen in the UN Sustainable Development Goals (SDGs) (UN, 2015). In terms of resource use, the concepts of renewability and natural resource sustainability are important. Sustainable use of bioresources can help to achieve many of the SDGs that in turn can improve social, economic and ecological conditions (Dietz et al., 2018). The European Bioeconomy Strategy includes: (1) ensuring food security; (2) managing natural resources' sustainability; and (3) reducing dependence on non-renewable resources, among its main five objectives, thus putting emphasis on sustainability (European Commission, 2012).

Some criteria for sustainability are scale-dependent, and we need to specify the spatial scale, from local to global. At the global level, the developed countries in the Global North import large quantities of energy, water, food, and ecosystem services in one form or another, and thus leave large ecological footprints, while developing countries in the Global South, with low levels of economic activity, contribute to smaller ecological footprints comparatively (Wackernagel et al., 2019). The production and consumption patterns of developed countries use copious resources provided by the developing countries, and they have large negative impacts on the environment or ecosystems in developing countries (Moran and Kanemoto, 2017). Such impacts include degradation of biodiversity and ecosystem services that in turn affects human well-being not only at local level but also on the global scale.

At the local scale, similar imbalances exist between urban and rural areas. Nearly half of the global population now live in cities, though many big cities cannot support their demands for water, energy, and food by themselves. Urbanization is expected to further increase, thereby putting more burden on the environment. In addition, most cities are sometimes free riders of ecosystem services. However, cities wield substantial economic power and rural areas depend on cities in terms of the economy and industrial products.

Such imbalances between developed and developing countries or urban and rural areas give rise to problems in terms of the methods used to evaluate sustainability, that is the spatial scale and criteria to evaluate sustainability. As mentioned above, urban and rural areas interact with each other in many ways. Hence, some man-made boundaries or administrative borders (like a municipality or a province) may not be an appropriate unit to evaluate sustainability. The appropriate units may differ according to the resources and ecosystems and their boundaries. While a watershed may be the correct unit to discuss water use sustainability, food and energy have other dimensions of resource supply and consumption. However, it is the administrative boundaries and the administrative agencies that scientists have to deal with in real-time conditions. The governance body is important since it regulates the production and consumption of resources, and it maintains the statistics necessary to evaluate sustainability. These are critical issues that researchers and practitioners face while evaluating sustainability. Thus, it is necessary to develop a system and standard criteria to evaluate sustainability. Such standard criteria can help regulating agencies to evaluate the performance and impacts of programmes or initiatives taken to promote development in a given area.

Several indices have been proposed to evaluate sustainability in the past, for example, the relationship between the Ecological Footprint (EF) and the Human Development Index (HDI) is frequently used to evaluate sustainability at the national or regional level (Wackernagel et al., 2019). However, if the same indices are applied at the sub-national scale, the results may not capture the reality and the statistics may not be correct, due to the large variations in the potential ability to provide resources and ecosystem services. The global ecological footprint does not capture the potential resource abundance at the local scale, which is a major drawback in estimation. Such a local consideration might be significant and important when we consider the interdependence between cities and rural areas or watersheds' sustainability.

The potential of rural and semi-urban areas to contribute goods and ecosystem services significantly to cities cannot be denied. These include agricultural products, food resources, clean water and energy, recreation, and so on. These goods and services are crucial to the economy in the rural areas as well as the cities they support and thus they are able to meet the objectives of relevant SDGs. In turn, the cities provide employment and incomes to rural people. The supply of goods and ecosystem services depends on how well the environment and natural resources are managed in a sustainable manner. In recent years, several governments have been trying to promote the management of natural resources through new approaches, such as bioeconomy or the green economy, as discussed in Chapter 1 of this book. These new approaches aim to promote sustainable resource management while generating income and employment with reduced impacts on the environment.

In the sustainability discussion, resources such as water, energy, and food are considered important since they are closely related to human life and the economy. In such discussions, many ecosystem services are often ignored and externalized with respect to the economy. However, ecosystem services are essential for human well-being, and must be maintained sustainably. Since 2010, several reports have been published specifying methods to quantify or evaluate ecosystem services for various ecosystems around the world (McVittie and Hussain, 2013). The methods to evaluate ecosystem services and changes to ecosystem services due to human pressures have been reported in some of the studies (Ministry of Environment Japan, 2016a). Recently, the ability to make these evaluations has significantly improved due to high accuracy and spatial resolution (Englund et al., 2017), though they are not yet deeply embedded in the discussions around sustainable societies. For example, in the Nordic countries, the contribution of bioeconomy projects to SDGs is evaluated based on five criteria including: the sustainable use of natural resources, technology innovation, environmental benefits, societal benefits, and the business model innovation (Nordic Cooperation, 2017). However, these methods cannot be generalized and used in all the contexts due to lack of proper data and monitoring of the ecosystem services, especially in developing countries.

In this chapter, we focus on the evaluation of sustainability at the local scale in Japan, for various resource uses, including some of the ecological resources (or ecosystem services) and the implications that it will have on promoting bioeconomy. The variation in sustainability among selected Japanese prefectures was analysed using indices for various resource uses in order to generate realistic values of local sustainability and give direction for the future. Sustainability indices and the potential of the rural areas to provide ecosystem services and support the economy of the cities, and vice versa, are important when showing cases to the policy-makers. This will help in attracting investment from governments and the private sector and thereby increase income and employment opportunities.

Assessing ecosystem services in Japan

Managing ecosystem services is an important part of sustainability. It relates to human well-being, including the provision of the materials necessary for human life, the regulation of the various ecological processes relating to security and environmental conditions around human life space, and sources or culture or mental health (Millennium Ecosystem Assessment, 2005). Ecosystems are renewable, and their management is crucial for a sustainable society.

With the importance of the ecosystem services concept growing, attempts to support quantification and evaluation of ecosystem services have advanced in recent years (TEEB, 2010). Ecosystem services and biodiversity as indicators have been used to evaluate various ecosystems worldwide (McVittie and Hussain, 2013), and in Japan as well (Ministry of Environment Japan, 2016a). Some ecosystem services (mostly provisioning and regulating services) have been quantified at the 1 km^2 grid scale using data from land-use or vegetation maps (Biodiversity Center of Japan, 2004), which has been repeatedly surveyed by the Ministry of Environment of Japan, and the data are freely available in geographical information systems (GIS) format. Such detailed information is becoming useful for developing concrete plans at the prefectural and municipality levels (Nelson et al., 2009).

A method to quantify some of the ecosystem services such as pollination (Taki et al., 2011) and agricultural production stability (Oguro et al., 2019), which otherwise had been rather difficult to quantify using the GIS, was also recently developed. In addition, techniques using big data and/or machine learning methods are also being applied to evaluate some of the cultural services (Aiba et al., 2019; Shibata et al., in preparation), which also suggests geographical imbalances in supply and demand using GIS.

Most ecological resources are renewable, though their usage should be considered in relation to the use of other resources, which may not always be renewable. Some resources, like energy, used to be a part of ecosystem services (firewood, charcoal, etc.) that rural areas provided to urban areas. However, now in Japan, the energy used is mostly from non-renewable sources, which is mostly imported from other countries. Recent trends reflect a change from non-renewable towards renewable resources, and the role of ecosystem services will thus be greater in the future. This may change the dependencies between urban and rural areas. Local sustainability largely relies on the production potential and manner of consumption of such essential resources (mostly water, energy, and food).

In this chapter, we will discuss sustainability of the prefectures in Japan using the available open source data and statistics, particularly on energy, food, water and selected ecosystem services for these resources, together with the quantified potential of ecosystem services in the area.

Variations in sustainability among prefectures in Japan

The variations in sustainability among prefectures in Japan are not well explained by plotting the Ecological Footprint (EF) versus the Human Development Index (HDI), since the range of EF and HDI does not vary much among the prefectures (4.06–5.24 for FE, and 0.91–0.97 for HDI) within the given global gradient (approximately 0.2–10.5 for EF, 0.35–0.95 for HDI). These indices tend to be convergent for local areas in a developed country, but might be divergent in developing countries, though we need more detailed analyses. It may also be related to the problem that EF is based on the uniform potential of ecosystems averaged over the world. There are variations in biological production and capacity of ecosystem services locally, though such variations are not considered. In fact, global EF does not consider the large amount of resource transfer between urban and rural areas in a given country.

Therefore, Principal Component Analysis (PCA) was performed using 32 parameters on resource use in 47 prefectures in Japan (Table 4.1) to understand the variation in resource use and sustainability among them. By matching many indices relating to production and consumption of resources together with population and economic activity, we were able to study the relationships between resource use and the characteristics of prefectures. This analysis may also provide information on how urbanization affects the sustainability of various resource uses.

Table 4.1 Factor loadings of parameters to PC 1–3 determined by PCA

Parameters	*PC1*	*PC2*	*PC3*	*Data source*
Population	–0,892	0,038	–0,106	Statistics Bureau of Japan (2016)
Population density	–0,919	0,063	0,208	Statistics Bureau of Japan (2016)
Per capita GDP	–0,392	–0,626	0,179	Cabinet Office of Japan (2016)
Per capita production capital	0,459	–0,508	0,379	Managi (2017)
Per capita human capital	0,257	–0,337	0,446	Managi (2017)
Per capita natural capital (market commodity)	0,900	–0,096	–0,125	Managi (2017)
Per capita natural capital (non-market commodity)	0,884	–0,110	–0,138	Managi (2017)
Ecological footprint (farmland)	–0,725	–0,603	–0,132	WWF Japan (2015)
Ecological footprint (pasture)	–0,723	–0,590	–0,139	WWF Japan (2015)
Ecological footprint (forest)	–0,701	–0,545	–0,330	WWF Japan (2015)
Ecological footprint (fishery)	–0,686	–0,609	–0,043	WWF Japan (2015)
Ecological footprint (built-up land)	–0,710	–0,598	–0,275	WWF Japan (2015)
Ecological foot print (carbon)	–0,510	–0,662	–0,386	WWF Japan (2015)

continued

Table 4.1 continued

Parameters	*PC1*	*PC2*	*PC3*	*Data source*
Per capita annual water use	0,859	−0,202	0,019	Water and Disaster Management Bureau (2019)
Per capita annual water use for household	0,540	−0,176	0,268	Water and Disaster Management Bureau (2019)
Per capita annual water use for industry	0,644	−0,362	0,164	Water and Disaster Management Bureau (2019)
Per capita annual water use for agriculture	0,750	−0,116	−0,142	Water and Disaster Management Bureau (2019)
Per capita CO_2 sequestration by ecosystems	0,345	−0,282	−0,029	Sasai et al. (2016)
Per capita CO_2 emission	0,103	−0,516	0,072	Ministry of Environment (2017)
Per capita electricity generation	0,269	−0,253	−0,019	Agency of Natural Resource and Energy (2018)
Per capita electricity demand	0,262	−0,649	0,388	Agency of Natural Resource and Energy (2018)
Per capita renewable energy supply	0,669	−0,279	−0,288	Institute for Sustainable Energy Politics and Itakura Laboratory (2018)
Per capita renewable energy potential (solar, electricity)	0,778	−0,418	−0,081	Ministry of Environment (2018)
Per capita renewable energy potential (wind, electricity)	0,795	−0,163	−0,248	Ministry of Environment (2018)
Per capita renewable energy potential (water, electricity)	0,782	−0,083	−0,056	Ministry of Environment (2018)
Per capita renewable energy potential (geothermal, electricity)	0,388	0,033	−0,537	Ministry of Environment (2018)
Per capita renewable energy potential (solar, heat)	0,746	−0,427	0,025	Ministry of Environment (2018)
Per capita renewable energy potential (geothermal, heat)	0,640	−0,390	0,195	Ministry of Environment (2018)

Parameters	*PC1*	*PC2*	*PC3*	*Data source*
Per capita agricultural production	0,217	0,120	–0,766	Ministry of Agriculture, Forestry and Fisheries (2016)
Self sustainability of food	0,617	0,072	–0,639	Ministry of Agriculture, Forestry and Fisheries (2016)
Self-sufficiency of field school for junior high school	0,378	0,243	0,347	Shibata et al. (in preparation)
Eigenvalue	12,697	4,711	2,708	
Proportion Explained	0,410	0,152	0,087	
Cumulative Proportion	0,410	0,562	0,649	

Most of the parameters used were collected from the statistics compiled by the Japanese government for recent years (see Table 4.1). These parameters cover the resources in terms of population, economic indices, such as GDP, human resources, and local production and consumption of resources (energy, water, food, and ecosystem services) in each prefecture. In addition to the government statistics, the parameters relating to Inclusive Wealth were obtained from Managi (2017).

Several statistics on ecosystem services were collected from reports issued by government and non-government organizations, as well as recent scientific reports. As for cultural ecosystem services, the variable 'field school' of junior high schools was chosen (Shibata et al., in preparation) as one of the recreation services, since it includes the data on human transportation among prefectures, which provides good information on the 'self-sufficiency' of a cultural service. This parameter is based on data collected by inquiring about the 'field schools' programme organized in junior high schools in every prefecture. Some of the junior high schools have 'field school' programmes, in which students participate in outdoor study tours of areas with unspoilt nature, such as forests, mountains, lakes or beaches. In the survey, the junior high schools were asked whether they have such a programme or not, and if they did, the destination of the tour was ascertained, followed by the cost of the programme. This parameter gives a ratio of the number of schools that have field programmes within a prefecture. A high value of the parameter means that the prefecture has the capacity to accommodate educational field trips for many schools. Some junior high schools in a big city like Tokyo have programmes to visit the southern islands located far away, while some schools choose nature-rich areas located within a short distance away. Using PCA, most parameters were calculated on a per capita basis, and all the parameters were applied after standardization.

The results of the PCA suggested the largest variation (PC1) was negatively related to the population density and population, and positively with per capita natural capital, and water use, suggesting some urban-rural gradient (Table 4.1). The variation along PC2 was negatively related to per capita GDP, ecological footprints (particularly carbon), per capita carbon emission and electricity demand, suggesting the gradient of economy and energy use. PC3 was negatively related to agricultural production and self-sustainability of food, suggesting the trends for production and consumption of agricultural products. The first PC explained 41 per cent of the variation and the cumulative proportion of explanation from PC1 to PC3 was 65 per cent.

To summarize, the prefectures in Japan could be characterized, first, by population or population density, second, by economy or energy usage, and then agriculture or food resources production and consumption. As for resource use, water is related mostly to PC1, while energy relates to PC2, and food relates to PC3. Since the urban-rural gradient is the most important factor affecting the variation in resource use in Japanese prefectures, the study focused on the sustainability of Japanese prefectures in terms of the relationship between the indices of sustainability and population density, which may provide suggestions on sustainability in the urban-rural gradient in this study.

Sustainable use of resources and ecosystem services in the urban-rural gradient

Here, we attempted to correlate several parameters relating to sustainability of the community with population density. The study analysed in detail the parameters in terms of self-sufficiency, considering the potential resource production and consumption of individual prefectures. We calculated the self-sufficiency of energy, water and food in association with ecosystem services in each prefecture. As for energy, we calculated the proportion of energy sourced from renewable energy instead of the self-sufficiency of energy (thus, ecosystem services may partly contribute to this). The self-sufficiency of water was defined as the proportion of per capita annual water use to annual rainfall amount in a prefecture. Also, for CO_2 emissions, we compared them to the carbon sequestration by ecosystems (one of ecosystem services) in each prefecture (Sasai et al., 2016).

Population density showed a positive correlation with per capita GDP, EF, and HDI (Figure 4.1 A, C, D), suggesting that highly populated areas tend to be economically rich and have better human development, though at the same time have greater impacts on the environment. When we examined the relationship between population density and per capita Inclusive Wealth, production and human capital had relatively vague negative relationships to population density, while natural capital had a significant negative correlation with population density (Figure 4.1 B). The

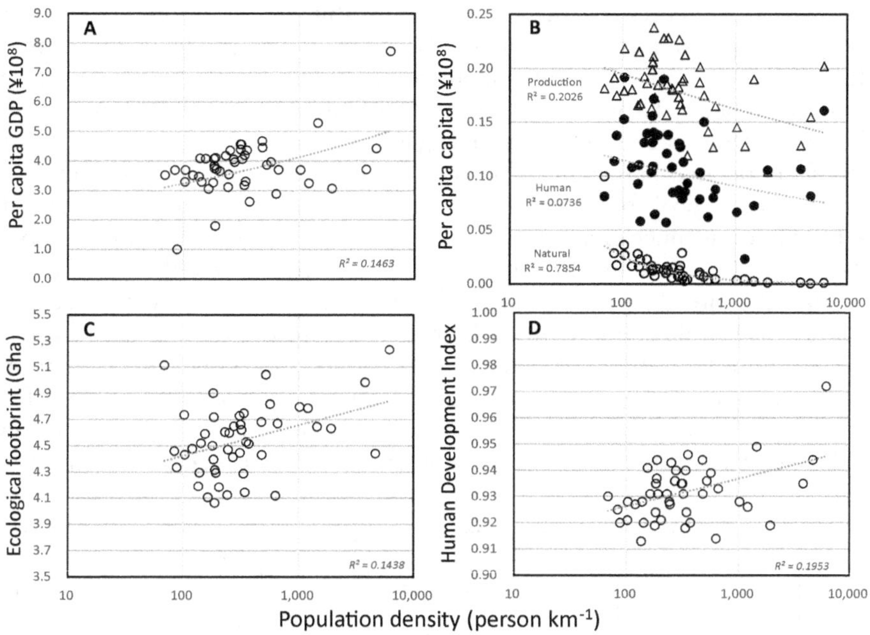

Figure 4.1 Correlations between parameters for sustainability and population density of prefectures in Japan.

Sources: A: Cabinet Office of Japan (2016); B: Managi (2017); C: WWF Japan (2015); D: Umemura (2003).

Notes

A: Per capita GDP; B: Production, human, and natural capitals; C: Ecological footprint; D: Human Development Index.

production and human capitals were much larger than natural capital, and the contribution of natural capital to the Inclusive Wealth was very small. However, we should be careful to note that a large portion of natural capital consists of forest timber, as other values of ecosystems were not included in the calculation (Managi, 2017), such that natural capital would be greater if we had included the values of other ecosystem services.

Per capita electricity demand did not show any clear relationship with population density (Figure 4.2 A), though the proportion of electricity provided by renewable energy was high in less densely populated prefectures (Figure 4.2 B). Similarly, per capita CO_2 emissions did not have a clear relationship with population density, as even some prefectures with less dense populations had very high emissions (Figure 4.2 C), probably because of low transport efficiency. Prefectures with less dense populations tended to use their own cars for transportation or more energy for agricultural production, and this leads to increases in per capita carbon emission.

However, the per capita CO_2 sequestration was clearly high in prefectures with less dense populations, because of the abundance of forest and other ecosystems absorbing CO_2 (Figure 4.2 D). These results partly confirmed that the prefectures with less dense populations tend to have smaller ecological footprints (Figure 4.1 C). If we discussed sustainability by self-sufficiency, i.e. the ratio of per capita emission to sequestration of CO_2, the tendency that densely populated prefectures have low sustainability will be more conspicuous than ecological footprint. However, we should note that the amount of sequestration is much smaller than that of emissions, suggesting it is far from above the CO_2 emissions than the ecosystem services generated.

Since the introduction of renewable energy is rising steadily, the proportion of renewable energy will be higher in the future, though only in prefectures with less dense populations. The prefectures with high population density did not show such potential, because of the scarcity of land for such applications. Prior to 1970, rural areas provided energy (firewood and charcoal, as provisioning ecosystem services) commercially to cities. Until

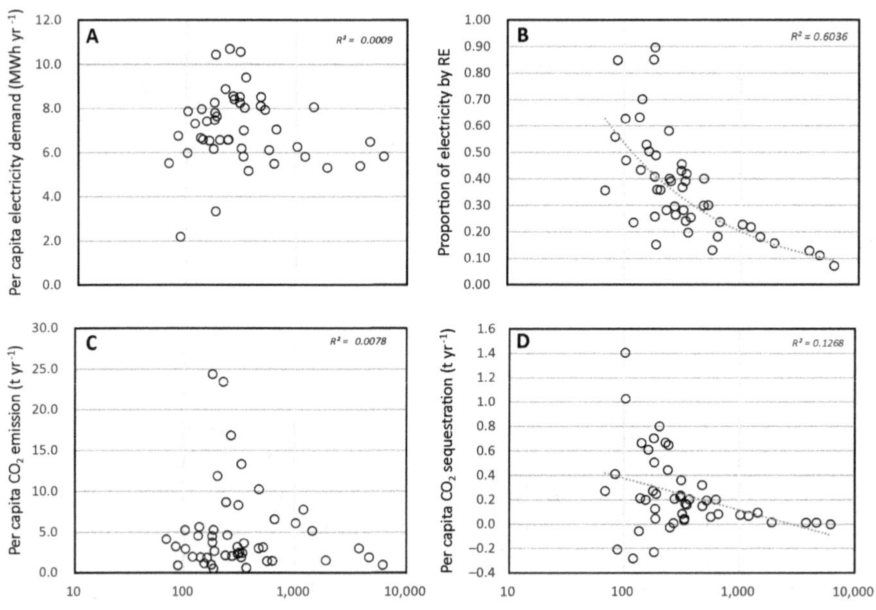

Figure 4.2 Correlations between parameters for energy use and population density of prefectures in Japan.

Sources: A, C: Ministry of Environment (2016); B: based on data from Institute for Sustainable Energy Politics & Itakura Laboratory (2018); D: Based on data from Oguro et al., in preparation.

Notes

A: Per capita electricity demand; B: Proportion of electricity provided by renewable energy; C: Per capita CO_2 emissions; D: Per capita carbon sequestration.

the 1970s, rural areas had renewable energy production that were much higher than their own consumption. However, after petrol and electricity became the main sources of energy for domestic and industrial use, biomass energy lost its commercial value. This means that rural areas used to provide energy (sourced from ecosystem service) to urban areas. The recent policies to use more renewable energy means that rural areas will again become important energy sources, but supplying new forms of renewable energy, such as solar, thermal, small-scale hydropower, and biomass-based energy.

As for water, prefectures with less dense populations had large amounts of water available for consumption (Figure 4.3 A), probably because of their limited water use for agriculture. This is largely because rice production, which is the dominant agriculture system in many prefectures, requires large amounts of irrigation water. However, including the water used for agriculture, per capita self-sufficiency of water (amount of annual water use/amount of annual precipitation in the area) was much higher in the prefectures with less dense populations (Figure 4.3 B).

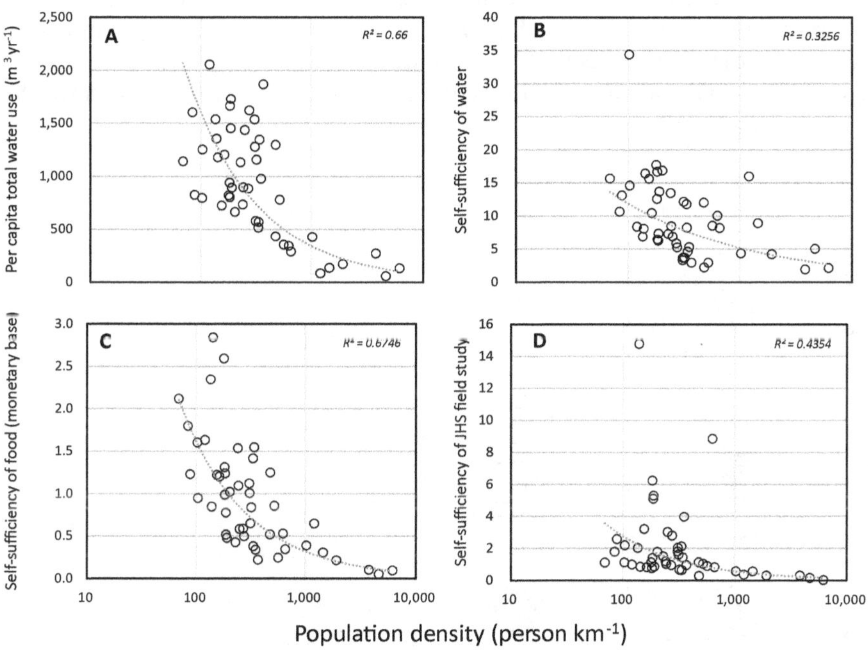

Figure 4.3 Correlations between the self-sufficiency of resources and population density of prefectures in Japan.

Sources: A: Water and Disaster Management Bureau (2019); C: Ministry of Agriculture, Forestry and Fisheries (2016); D: Shibata et al., in preparation.

Notes

A: per capita water use; B: Self-sufficiency of water use; C Self-sufficiency of food; D: Self-sufficiency of recreation

These prefectures also had high self-sufficiency in food (Figure 4.3 C), with large agricultural areas. The self-sufficiency of Japan was 35–40 per cent (in calories) or 65–70 per cent (monetary base), though many prefectures with less dense population excessed 100 per cent (monetary base), while the prefectures with dense populations were sometimes lower than 10 per cent, suggesting the difference in land use. Self-sufficiency of some cultural services was also high in the prefectures with less dense populations (visits to field school in this case, Figure 4.3 D). Prefectures with less dense populations had more lands rich in natural ecosystems. The junior high school was biased to select places with such rich nature (existence of lakes, seaside, mountain, forests, etc.) as the destinations for field trips (Shibata et al., in preparation).

If we look at the correlation between the parameters of self-sufficiency, most of them significantly correlated with each other (Figure 4.4), with less weak correlations between cultural services and others. Thus, the prefectures with less dense populations had high per capita consumption of some resources (water, energy), though they also had higher self-sufficiency for most of the resources they consumed. That is probably because of their low population density linked to having higher abundance of land and ecosystems than the prefectures with denser populations. These prefectures support high natural capitals and self-sufficiency of most of the renewable resources in the area.

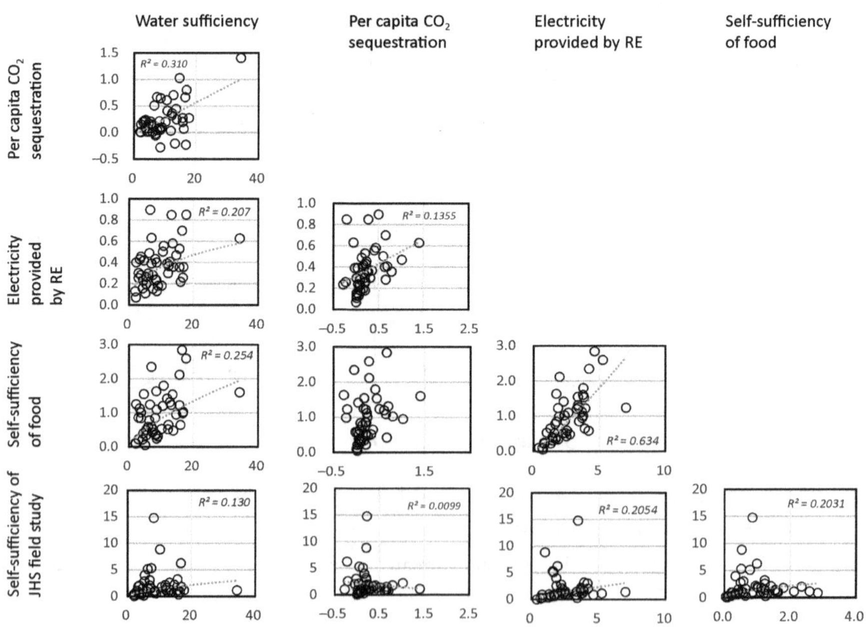

Figure 4.4 Links between the parameters for self-sufficiency of several resources.

Thus, areas with less dense populations (rural areas) tend to achieve a more sustainable society than denser populated areas (urban areas). Though rural areas also tend to be less well-off economically, and this could cause the population flow from rural areas to cities in Japan. The population imbalance between cities and rural areas will be enhanced in the future, though the self-sustainability of resources will be even higher in rural areas. In addition, cities are sometimes free riders in terms of ecosystem services such as energy, water, CO_2 sequestration, and some disaster risk reductions. Therefore, if cities could pay for these ecosystem services, it might lead to improvements in the imbalance between the economy and the population density between cities and rural areas.

Some PES (payment for ecosystem services) mechanisms are in operation in Japan. For example, a prefectural tax called the 'forest environmental tax' has been operating in more than 30 prefectures since 2007. The tax collected in a prefecture could be used for the management of upstream forests (Okada et al., 2007). Currently, foresters are subsidised to improve ecosystem services in plantations, but not much money is provided for ecosystem services in general. This system also has some problems in that the tax covers only the forest inside the prefecture, and is not applicable to the upstream forests beyond the prefectural range. Therefore, we should consider these challenges, such as the spatial scales and target ecosystem services to improve in the future.

Certification of sustainability is another way to bring the cost of managing ecosystem services from cities to rural areas. The proportion of the areas certified as sustainable forest management has been increasing in recent years, though the system has not been working very well in Japan (Nagaike, 2014). Because the forest owners in Japan operate on a small scale, it is hard to estimate and pay the cost. Also, the yarding timbers and labour for reforestation cost too much because of the steep topography, which makes it difficult to operate forest machinery, and this in turn increases the cost of management and maintaining ecosystem services. The premier price of certified timber is not sufficiently higher than non-certified timber, and hence does not give much motivation to the forest growers (Yamamoto et al., 2014). In addition, the value for certified timber together with the carbon market has not yet taken off in Japan. There is a need to resolve these challenges and develop good mechanisms for the PES mechanisms in Japan.

The concept of Inclusive Wealth is a way to change economic behaviour to improve the situation (Ikeda et al., 2017). Evaluation of natural capital could contribute to enhance the sustainable use of renewable resources. However, at present, the calculation of natural capital does not include capital relating to marine ecosystems. Even for terrestrial ecosystems, the timber production occupies most of the capital calculation. If we include the values of the other ecosystem services, the importance of natural capital will be greater, and contribute to increase the awareness of sustainable resource use.

Sustainability assessment of prefectures in Japan

The simple analyses performed in this study suggest several points about local sustainability. First, we can now access resource databases that enable analyses of the patterns of production and consumption of multiple resources simultaneously. Data on ecosystem services or ecological resources, which until now have not been quantified sufficiently, are accessible to the scientific community. The outputs from the analyses provide indications on human well-being rather than the resources that have been important for economic activity. If the measure of such human well-being is estimated in an appropriate way, our behaviour in terms of the use resources might be transformed. Therefore, we should not only have discussions on resources for commercial use, but also combine them with discussions on such resources that are important for sustainable human well-being. Looking at individual resources alone will help in partial optimization, rather, we need to consider the connections between multiple resources, including ecological resources. As seen in this study, ecosystem services are important indicators and tell us about the health of the ecosystem and how we need to change or adjust the production and consumption patterns to tread a sustainable development path.

Our analyses also made clear that prefectures with dense populations are economically rich with good human development, though they consume large per capita resources, and thus have high ecological footprints. Even though the prefectures with less dense populations consume larger amounts of water (for agriculture) and energy (for transport), and emit more CO_2, they also have higher self-sufficiency in terms of water, proportion of renewable energy, and per capita CO_2 sequestration. The local environment and ecosystems support their high self-sufficiency. In contrast, the prefectures with dense populations cannot support their needs by themselves, because of scarcity of land and ecosystems, and these rely more heavily on the externally provided resources.

These facts suggest that the local ecological footprint (ratio of resource consumption to local potential of renewable resources), rather than the global ecological footprint, is highly relevant and applicable to evaluate local sustainability. Local sustainability has a direct impact on urban areas, since these days cities receive most of their resources from outside, either from the rural areas in the region or from other countries. However, due environmental costs to produce or manage these resources elsewhere are not paid by cities. Ecological resources are typical examples, for example, freshwater resources, wood products, agriculture produce and tourism. Rural areas provide many ecosystem services to cities, though the sustainable supply of such ecosystem services requires a management cost. Mechanisms should be in place to ensure that such management costs are fully covered by the people who receive these resources. In this context, the relevance of instruments such as the PES becomes quite important, to ensure sustainable development in both cities and rural areas. If we evaluate using the local (not global) ecological footprints as the criterion, many prefectures in Japan seem very close to the goal of sustainability and achieve both low (local) EF and high HDI. Thus, some mechanism by which sustainable cities can achieve low EF seems critical to promote effective interactions with rural areas. Sustainable development is an important aspect of the bioeconomy approach, which sees beyond the area where production happens, thus establishing a link along the entire value chain including the consumers and the footprint created all along the chain.

There are still some issues to be addressed in order to achieve sustainable communities. First, the evaluation and analyses of ecosystem services are not yet adequate for the purpose. There is a need for some more studies to develop robust evaluation methods, including cultural ecosystem services. The connections between the ecosystem and other services should be analysed, together with cross-scale considerations. The contributions of ecosystem services to happiness, human well-being, and/or quality of life should also be quantified (Anand, 2016).

The indices and their potential to be applied to larger spatial scales are a challenge. For example, a watershed could be a convenient unit to discuss the sustainable use of water resources. An indicator developed within the watershed, for example, water use efficiency, could be applied to define that watershed scale. Many statistics can be accessed from local governance units such as the municipality or prefecture, though they usually do not have high independence in terms of resource use. If we want to discuss the sustainability in a city, we should include the rural areas with which they closely interact to get a holistic picture. We should note that every resource has its own geographical range of interaction, which should be considered while estimating the ecological footprint, an important index for sustainable bioeconomy.

The instruments developed should be specific and enable efficient and fair use of resources. The ability to document the imbalance between cities and rural areas, and/or developed and developing countries, will become more accurate in the future with new technologies and better access to data. Some instruments, such as the PES or other necessary measures should be developed and put in practice to avoid free riders. This may include developing better governance systems considering the urban-rural interdependency in the new bioeconomy initiatives.

Conclusion

In this chapter we attempted to analyse the resource sustainability of Japanese prefectures and proposed some possible criteria to evaluate local sustainability indicators. One of the key messages is that the ecological footprint values measured at the global scale may not be appropriate to discuss local sustainability challenges, rather 'local footprint' measures based on the self-sufficiency of the area could be appropriate. Experiences from Japan in this context could be relevant for other countries and may give new insights for the local governments to promote bioeconomy development. Cities are increasingly becoming dependent on resources from rural areas including tourism, experiencing nature and local food. It will be therefore difficult to attain sustainable resource management alone. Active partnerships are necessary where relevant stakeholders from both rural and urban areas can cooperate and manage resources, which have a huge potential to build a sustainable society even in Japan.

References

Agency of Natural Resources and Energy (2018) 'Electricity demand of prefectures 2017'. Available at: www.mlit.go.jp/mizukokudo/mizsei/mizukokudo_mizsei_fr2_000020.html

Aiba, M., Shibata, R., Oguro, M. and Nakashizuka, T. (2019) 'The seasonal and scale-dependent associations between vegetation quality and hiking activities as a recreation service', *Sustainability Science*, vol. 14, no. 1, pp. 119–129.

Anand, P. (2016) 'Happiness, well-being and human development: The case for subjective measures', UNDP Human Development Report, Background Paper. Available at: hdr.undp.org/en/content/happiness-well-being-and-human-development

Biodiversity Center of Japan (2004) 'Natural environmental survey'. Available at: www.biodic.go.jp/trialSystem/top_en.html

Cabinet Office of Japan (2016) 'Prefectural economic accounts during 2006–2015'. Available at: www.esri.cao.go.jp/index-e.html

Dietz, T., Börner, J., Förster, J.J. and von Braun, J. (2018) 'Governance of the bioeconomy: A global comparative study of national bioeconomy strategies', ZEF Discussion Paper. Bonn: ZEF.

Englund, O., Berndes, G. and Christel, C. (2017) 'How to analyse ecosystem services in landscapes: A systematic review', *Ecological Indicators*, vol. 73, pp. 492–504.

EC (European Commission) (2012) *Review of the 2012 European Bioeconomy Strategy*. Brussels: EU.

Ikeda, S., Tamaki, T., Nakamura, H. and Managi, S. (2017) 'Inclusive wealth of regions: The case of Japan', *Sustainability Science*, vol. 12, pp. 991–1006.

Institute for Sustainable Energy Politics and Itakura Laboratory (2018) 'Sustainable zone 2018 report'. Available at: www.isep.or.jp/archives/library/10867

Managi, S. (2017) *Evaluation of Wealth*. Japan: Chuo Keizai Sha.

McVittie, A. and Hussain, S.S. (2013) *The Economics of Ecosystems and Biodiversity: Valuation Database Manual*. Edinburgh: Sustainable Ecosystems Team at Scotland's Rural College.

Millennium Ecosystem Assessment (2005) *Ecosystems and Human Well-being: Synthesis*. Washington, DC: Island Press.

Ministry of Agriculture, Forestry and Fisheries (2016a) 'Agricultural production and income 2015'. Available at: www.maff.go.jp/j/tokei/kouhyou/nougyou_sansyutu/index.html#y1

Ministry of Agriculture, Forestry and Fisheries (2016b) 'Self-sufficiency of food 2015'. Available at: www.maff.go.jp/j/press/kanbo/anpo/attach/pdf/180808-2.pdf

Ministry of Environment, Japan (2016a) 'Japan Biodiversity Outlook 2 (JBO2)'. Available at: www.env.go.jp/nature/biodic/jbo2.html

Ministry of Environment, Japan (2016b) 'Greenhouse gas emissions 2015'. Available at: www.env.go.jp/earth/ondanka/ghg-mrv/emissions/results/index.html

Ministry of Environment, Japan (2017) 'Potential map for renewable energy'. Available at: www.env.go.jp/earth/ondanka/ghg-mrv/emissions/results/index.html

Moran, D. and Kanemoto, K. (2017) 'Identifying species threat hotspots from global supply chains'. Available at: www.nature.com/natecolevol

Nagaike, T. (2014) 'A review of the current status and perspectives of forest certification researches with special reference to conserving biological diversity in forest ecosystems', *Journal of Japanese Forest Society* vol. 96, pp. 267–273.

Nelson, E., Mendoza, G., Regetz, J., Polasky, S., Tallis, H., Cameron, D.R., Chan, K.M.A., et al. (2009) 'Modeling multiple ecosystem services, biodiversity conservation, commodity production, and tradeoffs at landscape scales', *Frontiers in Ecology and the Environment*, vol. 7, no. 1, pp. 4–11.

Nordic Cooperation (2017) '25 Nordic bioeconomy cases for sustainable change'. Available at: www.norden.org/en/news/25-nordic-bioeconomy-cases-sustainable-change

Oguro, M., Taki, H., Konuma, A., Uno, M. and Nakashizuka, T. (2019) 'Importance of national or regional specificity in the relationship between pollinator dependence and production stability', *Sustainability Science*, vol. 14, pp. 139–146.

Okada, K., Okada, S. and Yui, M. (2007) 'Process of policymaking in forest environment tax: A study on the rest tax Advisory Committee of the Iwate Prefecture', *Tohoku Journal of Forest Science*, vol. 12, pp. 1 –11.

Sasai, T.H., Obikawa, K., Murakami, S., Kato, T., Matsunaga, R. and Nemani, R. (2016) 'Estimation of net ecosystem production in Asia using the diagnostic-type ecosystem model with a 10 m grid-scale resolution', *Journal of Geophysical Research Biogeoscience*, 121, doi:10.1002/2015JG003157.

Shibata, R., Aiba, M., Oguro, M. and Nakashizuka, T. (in preparation) 'Influence of social-ecological attributes of landscapes on flow, demand and value of educational ecosystem service'.

Statistics Bureau of Japan (2014) 'Consumer Price Index 2013'. Available at: www.stat.go.jp/data/kokusei/2015/kekka.html

Statistics Bureau of Japan (2015) 'Family Income and Expenditure Survey 2014' (in Japanese). Available at: www.stat.go.jp/data/kokusei/2015/kekka.html

Statistics Bureau of Japan (2016) 'Search Statistics Surveys and Data 2015' (in Japanese). Available at: www.stat.go.jp/data/kokusei/2015/kekka.html

Statistics Bureau of Japan (2018) 'Statistics of Japan (e-Stat) 2017'. Available at: www.stat.go.jp/data/k-sugata/index.html

Taki, H., Yamaura, Y., Okabe, K. and Maeto, K. (2011) 'Plantation vs. natural forest: Matrix quality determines pollinator abundance in crop fields', *Scientific Reports*, vol. 1, pp. 132, doi: 10.1038/srep00132.

TEEB (2010) 'Ecological and economic foundations'. Available at: www.teebweb.org/our-publications/teeb-study-reports/ecological-and-economic-foundations/

Umemura, T. (2003) 'An analysis of the living standards of Okinawa Prefecture using Human Development Index (HDI) and Gender-Related Development Index (GDI)', *Ryukyu University Economic Review*, vol. 66, pp. 19–40.

UN (United Nations) (2015) 'Sustainable Development Goals'. Available at: www. un.org/sustainabledevelopment/sustainable-development-goals/

Wackernagel, M., Lin, D., Evans, M., Hanscom, L. and Raven, P. (2019) 'Defying the footprint oracle: Implications of country resource trends', *Sustainability*, vol. 11, no. 7, p. 2164.

Water and Disaster Management Bureau (2019) 'Water Resource Report 2018'. Available at: www.mlit.go.jp/mizukokudo/mizsei/mizukokudo_mizsei_fr2_000020.html

WWF Japan (2015) *Ecological Footprint of Japan 2014*. Tokyo: WWF.

Yamamoto, Y., Takeuchi, K. and Shinkuma, T. (2014) 'Is there a price premium for certified wood? Empirical evidence from log auction data in Japan Forest', *Policy and Economics*, vol. 38, pp. 168–172.

5 Transition towards bio-based economy in small-scale agriculture in Sub-Saharan Africa through sustainable intensification

Mehreteab Tesfai, Giacomo Branca, Luca Cacchiarelli, Chiara Perelli and Udaya Sekhar Nagothu

Introduction

The agricultural sector has great potential to produce adequate food and biomass while contributing to economic growth, if managed properly (FAO, 2017). In Sub-Saharan Africa (SSA), the agricultural sector provides, on average, more than 10 per cent of the export revenues (Sasson, 2012), and accounts for a significant proportion of the gross domestic product (AGRA, 2016). The sector also contributes to socio-economic development through job creation and is an important source of raw materials for many other rural sectors, including bio-energy and livestock. About 60 per cent of the rural population in SSA depend predominantly on small-scale agriculture for their livelihoods (AGRA, 2013). In many SSA regions, small-scale agriculture is predominantly subsistence-oriented, characterized by low levels of inputs and low outputs, predominantly carried out in small parcels of land and based on a mix of indigenous and scientific knowledge. Today, smallholders are at the centre of the challenges associated with population growth, urbanization, climate change, resource scarcity, food and nutrition insecurity and are poorly connected to markets while having few opportunities for value addition to their production and post-production activities (Gollin, 2014).

African agriculture is becoming increasingly dependent on fossil inputs mainly due to increasing use of synthetic fertilizers and other agrochemicals. The production model is largely 'linear' in structure where inputs are used to produce outputs (in the form of food, feed, fibre) and then dispose of the rest (i.e. the take→make→use→dispose one-way model that is predominant). In addition, losses during harvest, transport, storage and processing also lead to losses of about 20–25 per cent, which can be reduced by improving the post-harvest facilities. Such a linear way of production results in wasted resources and environmentally damaging outputs. By increasing the production efficiency, recycling and re-using the materials produced, the pressure on the environment and natural resource bases can be minimized.

The bio-based economy (BBE) is one of the possible solutions for African agriculture to become more efficient in its resource use (Ward et al., 2016). In agriculture, this implies the production of agricultural commodities by using renewable natural resources while reducing waste, with the aim of increasing resilience and/or sustainability. BBE offers solutions in moving towards a sustainable system, involving changes in food production, consumption, and residue management (Jurgilevich et al., 2016). Schroeder, Anggraeni and Weber (2018) have suggested that BBE can directly help to meet some of the UN Sustainable Development Goals (SDGs) (UN, 2015) (Table 5.1).

Many of the bio-based economy analyses have mainly focused on isolated and, to a large extent, technological transition pathways (Bennich et al., 2018). Examining the entire agri-food value chains from the 'bio-based economy' perspective may reveal a range of drivers and barriers to transform small-scale agriculture into bio-based economy (Ward et al., 2016). Sustainable intensification (SI) options can provide the pathway to BBE approach that can transform linear farming systems into circular models. According to Pretty, Toulmin and Williams (2011), 'SI can produce more outputs with more efficient use of all inputs, through various intensification processes on durable basis while reducing environmental damage and building resilience, natural capital and the flow of environmental services.' Sustainable intensification of agricultural practices and value-addition of the bio-based products could transform the existing farming systems into more

Table 5.1 Bio-based economy contributions to most relevant SDGs

SDGs (most relevant)	*References*
SDG 2: End hunger, achieve food security and improved nutrition, and promote sustainable agriculture	Juerges and Hansjürgens, (2016)
SDG 6: Ensure access to water and sanitation for all	BMEL (2014)
SDG 8: Promote inclusive and sustainable economic growth, employment and decent work for all	Kalmykova et al. (2018)
SDG 12: Ensure responsible consumption and production	Hoekstra and Wiedmann (2014)
SDG 13 Take urgent action to combat climate change and its impacts	UN, (2015)
SDG 15 Sustainably manage forests, combat desertification, halt and reverse land degradation, halt biodiversity loss	Juerges and Hansjürgens, (2016)

sustainable ways of agricultural production. For example, the incorporation of legumes with maize-based systems either as intercrop or by crop rotation provides a simple and effective strategy towards SI in smallholder farming systems in SSA (Thierfelder et al., 2012). Intercropping with legumes can generate multiple benefits to the soil by fixing nitrogen, produce diverse outputs on the farm and improve human nutrition when consumed by households. However, there are risks associated along the value chain (VC) in a new system. There is a lack of adequate knowledge about the drivers, barriers and conditions necessary to transform linear agriculture into a circular BBE model. This is particularly true in the African context where the BBE approach is relatively new, the majority of farmers are smallholders with limited capacity to take risk and invest in new approaches without being assured of the outcomes.

The aims of this chapter are: (1) to review and analyse the potential of sustainable intensification (SI) and demonstrate the importance of mapping the agricultural VCs; and (2) to discuss how to promote BBE in the small-scale agriculture sector of SSA. In this chapter, experiences from the case study of the maize-legume cropping systems in Malawi, which is a good example of a simple agronomic system capable of delivering economic, environmental and social benefits (Tesfai et al., 2019) are shared.

Following the introduction, the rest of this chapter is structured as follows. The conceptual framework is presented in the next section, followed by the case-study area description. The results of the study are then presented, which includes the BBE technology options available, the main drivers and barriers to transform the small-scale agriculture into a BBE. The next section discusses the transition pathways towards a BBE in the Malawian context. Finally, the chapter provides some recommendation and concluding remarks.

Conceptual framework

The three main components of SI are: (1) increased production, income, nutrition or other returns; (2) reducing greenhouse gas emissions; and (3) strengthening resilience and reducing environmental impacts. In Table 5.2, the corresponding sustainability dimensions (i.e. environmental, economic, and social) of the SI components are shown. A balance of the three sustainability dimensions, the environment, the economy and society is desired in a system, that needs to be integrated (Table 5.2). SI should thus help to develop an agricultural production system that is socially acceptable, economically feasible and environmentally friendly to producers and to society at large (FAO, 2011). This may be easier said than done, and it will not be an easy path to achieve this, given the current situation prevailing in the agricultural sector in SSA.

Table 5.2 Principles of sustainable intensification and corresponding sustainable dimensions

SI components	*Principles*	*Sustainable dimensions*		
		Environmental	*Economic*	*Social*
Increased production, income, nutrition or other returns	On the same amount or less land and water	X		
	With efficient and prudent use of inputs		X	
	Using productive use of knowledge and capacity			X
Minimizing greenhouse gas emissions	While increasing natural capital and flow of ecosystem services	X	X	X
	Reduce impact on forests via alternative energy sources	X	X	
Strengthening resilience and reducing environmental impacts	Through innovative technologies and processes	X		
	Minimizing use of technologies or inputs	X	X	

Since production alone will not solve the current and future food, feed, and fibre demands, the entire agri-food value chain should be considered by looking at the 3R principles of *R*educing inputs, boosting *R*euse and *R*ecycling bio-residues from the agricultural production, storage, processing, marketing and consumption phases.

The agricultural food VC is defined as 'the full range of farms and firms with their successive coordinated value-adding activities along the chain that produce agricultural raw materials and transform them into products that are sold to consumers and disposed/or recycled after use'.

An integrated and renewable bioeconomy-based mixed farming systems that could be relevant for smallholders is shown in Figure 5.1. There are several types of biomass that can have valuable applications in the BBE including crops, grass, woody biomass, aquatic biomass, straw, animal fat and manure. For example, livestock manure may be viewed by some as 'waste', but it is a valuable resource for agricultural land due to its potential

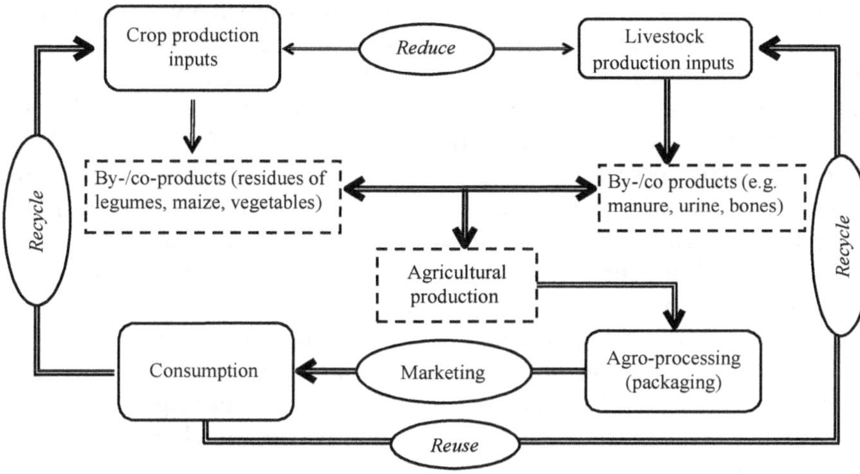

Figure 5.1 Bio-based economy of mixed farming systems along the entire value chain.

to increase soil nutrients and fertility. Farm manure is commonly used by smallholders who own livestock in SSA. Some farmers even sell manure to others, thus making some additional income. However, composting techniques and time of manure application are critical factors that smallholders need to be aware of and this would require training. Farmers also use crop residues such as straw, leaves and roots to maintain soil organic matter levels, improve the soil moisture-holding capacity and enhance the soil structure.

Utilization of animal manure and crop residues along the supply chain is expected to reduce the external supply of chemical inputs, increase the reuse of by-products and improve recycling. This will help to reduce the net amount of fossil fuels and mineral fertilizers used on the farm to produce food, feed and fibre. Within the BBE framework, outputs from a farm should ideally become inputs to the same farm and vice versa. Integration of BBE principles within sustainable supply chain management can provide socio-economic and environmental benefits (Genovese et al., 2015).

Case study description

A case study of the maize-legume system in the Dedza district of Malawi was used to analyse the potential of SI and its contributions to promote a bio-based economy. Malawi is a small landlocked country located in the southern part of Africa with a population of 15 million, and most of them are dependent on agriculture for their livelihood (World Bank,

2014). They are mostly smallholder farmers who own less than 2 hectares of land. Approximately half of the calories consumed by rural households are derived from their own food production (Dorward and Chirwa, 2011). Maize is the dominant staple crop in the country, covering over 60 per cent of the arable land and constituting about 70 per cent of calories consumed (Ellis and Manda, 2012). Most farmers lack the resources to invest in improving the productivity of their farm, or to undertake value addition to their produce. Low productivity, lack of stable markets and little economic diversification make farmers vulnerable and food-insecure. This is leading to a continuous migration towards urban areas, especially among the younger generation in Malawi. We are observing a similar outmigration in other SSA countries. In many locations, unsustainable practices have serious impacts on the environment, such as land/soil degradation that further aggravates the vulnerability of the farming community due to climate change, extreme weather and crop pest outbreaks.

As part of the ongoing InnovAfrica project (www.innovafrica.eu), maize-legume cropping systems were introduced in Dedza district in Malawi. Improved maize varieties that are rich in nutrition (e.g. orange maize) are now being grown together with different types of grain legumes (e.g. pigeon peas, soybeans, groundnuts) in rotation or intercropping. The trials were farmer community-led, and tested and/upscaled in different farmers' fields/plots (n = 72). Farmers tested the conservation agriculture practices against the conventional practice of monocropping. Women constituted a significant percentage of the new adopters in this case. The crops were grown on raised and flat-bed ridges, fertilized with compost that is locally made using different methods as described in the next section of this chapter. The data from the first two cropping seasons (2018–2019) showed that intercropping of maize-legume systems has produced a higher grain yield than monocropping.

A mapping exercise was simultaneously done to identify the VC actors operating at various stages: *the existing links*, *the product flows and the enabling environment*. Figure 5.2 shows the maize-legume value chain mapped in Dedza district, from the phase of pre-production (i.e. resources used as inputs to on-farm production process) to post-production (i.e. post-harvest-processing-marketing-consumption-reuse/recycle).

Besides farmers (producers) there are several other actors (e.g. traders, processors, consumers) and institutions (e.g. markets, government, producers' organizations, rural service providers, rural finance entities) who influence the VC. The main characteristics and VC links are described below, together with the role that VC actors can play in increasing resource use efficiency through re-using and recycling options.

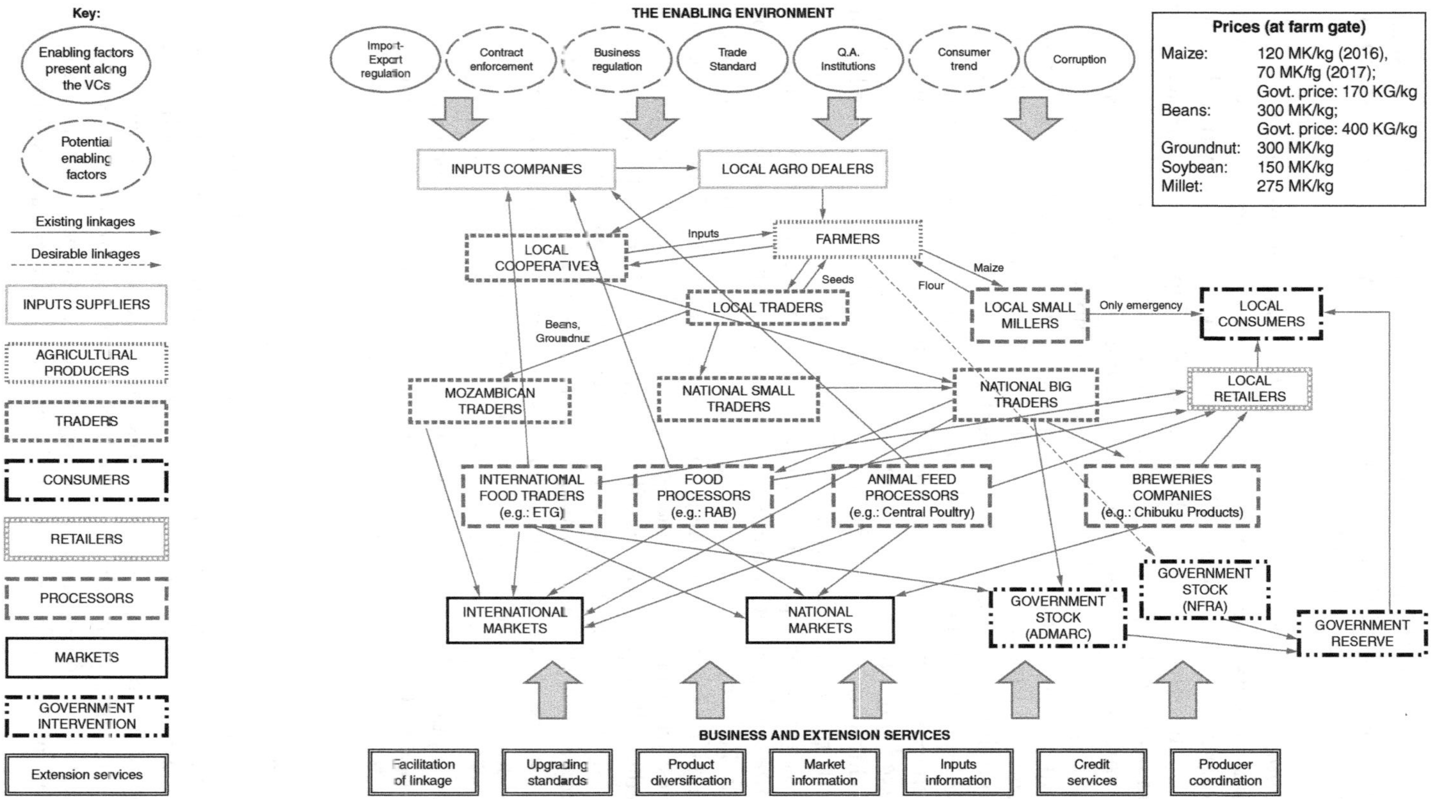

Figure 5.2 Maize-legume value chain map in Dedza district in Malawi.

Input providers

This phase of the VC includes the provision of farming inputs to smallholders. It was found that three input provider typologies operate in the case-study area as described below.

1 *Companies that produce and market their own inputs*: There are a few input dealers operating in the market, for example, the company *Exagris Africa Limited* (Malawi branch) produces certified seeds through grower schemes. Farmers receive inputs on a contract basis and produce dry paprika, groundnuts and maize seeds.
2 *Companies exclusively engaged in inputs trading*: Companies such as the *Export Trading Group (ETG)* sell inputs to farmers purchased from national and multinational companies through the company's agents and local agro-dealers. The ETG mostly sells seeds (of maize, groundnuts, soya beans and vegetables), fertilizers, pesticides and herbicides to both farmers and cooperatives and works in collaboration with extension agents who provide technical assistance to farmers.
3 *Mixed producer-trader companies*, which trade their own-branded products together with different categories of inputs purchased in the market. For example, *Peacock Enterprise* is a private entity that sells seeds and other inputs and provides technical assistance to farmers. The maize seed market is highly concentrated, with a high share – 80 per cent kept by multinational companies (e.g. Monsanto, SeedCo and Pannar) while the remaining 20 per cent is held by 25 local companies. Sales arrangements include payments in cash and, less frequently, on credit. Selling prices are based on other vendors' prices, buyers' demand, costs, location, time of year. Farmers can get discounted prices for bulk purchases or through – mainly informal – agreements with preferred customers or farmers' organizations. About 35 per cent of farmers in the study area can get access to the government-supported Farm Input Subsidy Programme (FISP) aimed at increasing access to seeds and fertilizer inputs. The other 65 per cent smallholders are dependent on the private sector to purchase fertilizers and seeds, which can be a challenge due to price fluctuations and quality variation.

Output market

On-farm production of maize, beans, groundnuts and soybeans is used both for self-consumption and commercial purposes. Farmers receive technical assistance from agro-dealers, extension agents and public media (mainly radio) but have limited access to services. Seven farmer cooperatives, with approximately 75 members each, operate in the area. They provide production coordination and storage services, facilitating market access to members, buying grains from their members and selling to big traders from the cities of Lilongwe and Blantyre in Malawi. Farmers pay an annual membership fee and can buy shares in the cooperative.

Output processing and value addition

Small-scale local millers (processing mainly maize produced for self-consumption), and large-scale millers owned by Malawian or international traders offer milling services (e.g. *Bakhresa Malawi Ltd*, located in Southern Malawi). Also, several companies producing food, feed and beer operate in the maize-legume value chain, such as:

- the *RAB Processors Limited* specializing in agro-processing, using food fortification technology to produce outputs to be used in malnutrition alleviation programmes in the country;
- the *Central Poultry* processes a diverse range of feeds and markets fresh and frozen chicken as well as commercial eggs;
- the *Chibuku Products Limited* is involved in several business activities, including manufacturing malt-based beverages, such as beer and liquors;
- the *Export Trading Group* processes soya beans into different products (soya pieces, soya milk and soya cooking oil).

These entities could play a useful role in helping smallholders grow quality maize with better nutrition (orange maize) or special maize for brewing that could fetch better prices.

Marketing

Product differentiation, processing and marketing at the smallholders' level in Malawi are almost non-existent. Farmers generally sell raw products mainly at harvest time and without any value addition. The price is simply decided by intermediaries as the farmers have limited bargaining power. This is one area where bio-based VC development can help farmers get a reasonable price for their products. On the contrary, large-scale farm enterprises do add value to their products, competing better than smallholders on the markets.

Traders

Three categories of traders operate in the study area: They include:

- *local traders*, who buy raw material from local smallholder farmers;
- *national traders*, who buy from local traders and sell to international traders and processors;
- international traders (e.g. the Export Trading Group, RAB Processors Ltd, Farmers World Ltd) which operate on the international market.

To a certain extent, maize and legumes form part of the traditional staple food consumption system. Local consumption is mainly based on locally grown crops and locally milled floor. Traded and processed products are sold in supermarkets in main cities.

Extension services

Market-related information is provided by different organizations and institutions. The National Smallholder Farmers' Association of Malawi (NASFAM Commercial) supports farming as a business. Its services include trading support, product development and commercialization, marketing strategies, wholesale and retail sales, financing programmes and input supply. The extension services provide information about markets and agricultural inputs. In Dedza district, there are several agricultural-oriented non-governmental organizations (NGOs) that provide technical assistance and training to the groups of people they work with. However, the farmers' participation in group activities like clubs at the village/area level is variable. This leads to a heterogeneous farmers' awareness and risk perception about market potentials. The credit and financial service providers include Opportunity Bank, Finca, MADEF, Seed Loans (provided by cooperatives) and Village Saving Loans. The main issue related to credit services is to make credit easily accessible to farmers, whether they can provide collateral or not. Through village saving and loans groups (VSLGs), members pay a monthly fee and make savings in the group where they can borrow money from. This system is becoming popular and allows farmers to take out a loan for agricultural and other activities.

Enabling environment

The climatic conditions in Malawi are unstable. The country has recently experienced two long droughts since 2015. To tackle internal food shortages, the government bans maize export at times when unfavourable climatic conditions affect agriculture production. Import-export regulation is particularly enforced on maize. Government price control on maize is aimed at protecting farmers from price fluctuations on the market. Nevertheless, good harvest years and the subsequent high maize supply might lead to lower prices in the local markets. Standards for trading vary according to the destination market and a price list based on grain quality categories exists. Quality control measures are rarely adopted by smallholder farmers. This is due to lack of awareness, capacity, facilities for quality checking, and lack of access to inputs and technology, which adversely affect production of high-quality output.

Sustainable intensification options

There are several simple solutions that can promote SI and help to improve the efficiency of the inputs used, promote re-use options, and/or recycle outputs from a farm, putting the principles of the 3Rs into practice. Some of them are simple, of low risk and easy to practise if farmers are made aware of the benefits of the practices. Examples of SI technologies introduced in the InnovAfrica project and practised by farmers in the case study area are discussed below:

Agricultural inputs

1 *Integration of legumes (with conservation agriculture)*: Legumes integration in the farming system spatially or temporally can reduce nitrogen fertilizer inputs to the soils, as legumes fix significant amounts of nitrogen in the soil (Vanlauwe et al., 2019). This is a low-cost intervention and is environmentally friendly, typically suitable for smallholder settings. Guidance on the optimal use of crop residues from legumes for soil improvements could be useful for smallholders as it helps to maintain soil organic matter levels. In smallholder settings of Malawi, the main component for livestock feed is often crop residues. Therefore, high quality feed residues from grain legumes could be a good option, as they clearly have an added value in the economics of the crop-livestock VCs. Livestock fed with legume crop residues have shown higher milk outputs. Legumes/beans when consumed by households have also shown better nutrition and health impacts according to several studies.

2 *Need-based input application/Precision farming (PF) methods*: The International Plant Nutrition Institute (IPNI) states:

> Precision farming allows us to carry out targeted application of nutrients at the right time with the right quantities, depending on the actual needs of the plant – something that traditional agriculture today tends to neglect. PF utilises the capabilities of information technology systems to optimise the application of agricultural inputs (e.g. fertiliser, agro-chemicals) by delivering the 4Rs.

The fertilizer 4Rs refer to the need to ensure farmers access to the *R*ight source of fertilizer for their crop and soil, apply it at the *R*ight time in relation to crop development and growth in the *R*ight place close to the plant and in the soil using the *R*ight method of application. Such practice ensures that the minimum resources needed are used in order to achieve optimum performance with minimal environmental impact (IPNI, 2012).

PF addresses the use of minimal levels of invested resources that is essential to achieve sustainable agricultural production (Ward et al., 2016). The

BBE approach may be more suitable, where upstream inputs (investment in fossil inputs, e.g. fuel, phosphorus, soil) are minimized using PF and downstream outputs (e.g. manure, crop residues) are 'circulated' possibly via SI pathways. One highly efficient and intrinsically sustainable approach of PF is the technique of *micro-dosing*. Micro-dosing minimizes the application of and over-reliance on inorganic fertilizer, improves nutrient use efficiency and protect against drought (The Montpellier Panel, 2013). For instance, farmers implementing micro-dosing in Niger, Mali and Burkina Faso have increased millet yields by over 50 per cent and improved soil water retention (ICRISAT, 2009). Micro-dosing is a simple but robust technology that can be operated and maintained locally and is suitable for operation at the farm level. Micro-dosing of fertilizer application is one relevant option for smallholders in Malawi. Simple PF techniques which do not require expensive infrastructure must be provided to smallholders, for example, mobile apps that could advise farmers on how to regulate their fertilizer and water inputs.

Re-use the agricultural output

Much of the agricultural waste during the production phase, typically described as by-products and/or co-products, is unavoidable. These products are composed of manure, crop residues, leaves, peels which are re-used for other purposes by applying various techniques. The common ones re-used by local farmers in Malawi are described below:

1 *Composting*: One of the technologies to facilitate the valorization of agricultural organic wastes is composting. Composting (a form of recycling) is a natural way of returning biological nutrients to the soil (Ellen MacArthur Foundation, 2013). Composting has its own benefits and drawbacks, such as *carbon sequestration*, *return of organic matter*, *nutrient recycling* but may also cause environmental damage if not properly managed (Carus and Dammen, 2018).

   Farmers in Malawi enrich their soils by adding compost to their fields. Five different types of composting methods are practised by the local farmers. These are locally called:

   (a) *chimato*: a mound-like structure with ingredients of animal manure, and household waste;
   (b) *pit composting*: a pit with a mixture of crop residues, manure from pigs/goats, and top soils harvested from nearby forest land;
   (c) *manyawa a bokashi*: composting materials with a mixture of pig manure, maize brans, top soils harvested from forest land, ashes, leaves of acacia trees, crop residues;
   (d) *mpanda*, also called fence composting, that is produced from mixture of manure, soils, and crop residues.

Participating farmers in the InnovAfrica project (www.innovafrica.eu) have stated that crop yields from the maize-legume trials increased significantly when using compost manure. Farmers showed a positive response to these simple and low-cost interventions by the project, which resulted in higher yields and diverse farm products A portion of the produce is consumed by local households, particularly legumes, which helps with better nutrition. This was evident from the statement by a woman farmer who said, 'Consuming orange maize with legumes is the secret of my better health even at the age of 75.'

2 *Application of animal manure*: In Malawi, animal manure is one of the common valuable resources used for soil improvement. Most farmers in the study area use farm manure to enrich their soils (Mungai et al., 2016). The manure is produced year-round but stored and used seasonally. The contents of manure in terms of nutrients and the amount produced vary depending on the type of animal and the animal husbandry/management practices. The use of animal manure as a source of nutrients reduces the quantity of mineral fertilizers required to be applied to the soils and this in turn contributes to the mitigation of greenhouse gases (GHGs) emissions. In addition, animal manure can also be used to produce biogas and the residual slurry (a by-product of biogas), which can be applied to fertilize soils and increase crop yields (see Chapter 10 in this volume). Farmers should manage the available manure on their farm effectively, otherwise, it can be a source of environmental pollution.

3 *Incorporation of crop residues*: Crop residues from above ground (e.g. leaves, stalks, broken seeds, etc.) and below ground parts (e.g. roots, stalks) are potential resources to be used for mulching, improving soil fertility, recycling of nutrients, reducing soil erosion, and conserving moisture in the soils. The majority of the farmers in the study area apply at least one of the three principles of conservation agriculture (ibid., 2016). For example, incorporation of crop residues (of legumes and maize) into the soils and/or covering the soil with mulching to conserve soil moisture. In a mixed farming system, judicious use and management of crop residues are vital to enhance sustainable farming systems (Bisen and Rahangdale, 2017). On the other hand, burning of the maize stalks and removal of crop residues reduce organic inputs, which can lead to loss of soil organic matter, insufficient nitrogen supply, ultimately decreasing the soil's water-holding capacity and crop productivity (Mungai et al., 2016). Moreover, burning of biomass, such as maize residues, emits GHGs, causes smoke and pollutes the environment (CEC, 2014).

Recycling the agricultural products

In agriculture, the pathways for products recycling take different forms including the gasification of crops; using crop stalks as feed, fertilizer or building materials; and making manure from livestock excrement and creating new value chains (Wang et al., 2016). However, resources such as crop residues and manures can remain without being 'circularized' but may be valorized to produce bioenergy/chemicals for the wider bioeconomy. One such example of *valorization of bio-based products* is the returning of nutrients to farms by composting (as mentioned above) that improves soil fertility and by growing different combination of nutritious crops over seasons (temporal) and at different place (spatial).

Barriers to adoption of sustainable intensification options

Technological barriers

Small-scale farming in the study area is mainly practised using indigenous knowledge that has been transferred from generation to generation. Although the system implements some sort of circularity of resources within the VC, by using agricultural wastes, residues and by-products, the BBE principles and practices are not mainstreamed in the farming system. Smallholders have limited access to innovation as information and communication technologies (ICT) are not readily accessible and/or not affordable by them. This is due to poor infrastructure and weak extension services that limit access to technology information. The implementation of innovative farm techniques can be facilitated by access to agricultural inputs (e.g. land, pesticides and improved seeds) and extension services, as well as by participation in farmers groups and organizations. Also, an awareness campaign of BBE principles and applications can help farmers and other stakeholders to access technologies. However, implementation of BBE practices in dry areas of SSA may be a challenge due to lack of agricultural biomass such as crop residues and manure. There is a high competition to use crop residues for fuel wood or construction material, for livestock (as a feed source) and as fertilizers for soils.

Market barriers

Agricultural marketing systems in Malawi are still rudimentary, and in some cases non-existent in rural areas. This is partly due to inadequate infrastructure for efficient agricultural marketing; limited access to and poor quality of marketing service provision; and policy incoherencies that negatively affect marketing. These constraints limit the incentives and ability of farmers to increase their participation in agricultural value chains for both domestic and export markets (MoAIW, 2016). Absence of market and transport facilities can discourage farmers in rural areas from making more effort to develop by-products such as feedstocks. High costs of pre-treatment, poor storage facilities and lack of transportation aggravate the challenges for the VC actors.

VC actors and smallholders are encountering problems associated with market information, outlets, prices, under the linear economy (LE) mode of production. Farmers' problems continue after harvesting the produce (e.g. low and unstable farm-gate prices), whereby they do not find good market prices to cover their production costs and other household expenses. Investments to introduce proper labelling and marketing communication to enhance the expected social, environmental and economic benefits accrued from the implementation of BBE options should be made at various stages of the chain, considering the time lag between costs' implementation and benefits gained. Consumers may be willing to pay a premium price for green(er) services/bio-based products provided that proper information is channelled to them. In this respect, the agri-food VC actors could be instrumental in connecting producers and consumers, for example, farmers markets where consumers can directly buy from farmers producing niche products, organic vegetables, or other by-products.

Policy/institutional barriers

At present, most African governments give little/no attention to promoting BBE, particularly when it comes to the agriculture sector. One of the reasons could be lack of awareness of BBE. Except for a few countries, there is no specific policy statement/instrument or legislation or institutions mandated to develop BBE. In the case of Malawi, the National Agriculture Policy (MoAIW, 2016) does not include any measures to promote BBE. The policy gives guidance on increasing production, productivity and real farm incomes (ibid.). Lack of policy support and lack of an institutional framework for BBE hamper the development and overall process of BBE transition in small-scale agriculture. Government support to smallholders in the form of credits, capacity building, training, skills development, creating enabling environment, is vital during the transition to BBE.

Smallholders are often reluctant to adopt new technologies and are sceptical of innovations. They may be unwilling to collaborate to introduce elements of circularity in their linear model. Sometimes, this could be due to increased need for labour, or lack of market for the products. Bio-based outputs and markets require social acceptance among the various value chain actors.

Transition pathways towards a bio-based economy

The fundamental drivers to BBE can be broadly categorized as *environmental* (e.g. environmental pollution as the LE aggravates adverse climate change impacts), *economic* (e.g. resource scarcity as the LE is wasteful and unsustainable) and *social* (e.g. lack of new knowledge/information because the LE functions as business as usual). The conditions to accelerate the willingness to adopt BBE arise from the pressure to reduce negative environmental impacts caused by LE (the current business-as-usual model). From the economic point of view, the willingness to adopt BBE stems from the need to increase resource use efficiency. These drivers and their associated benefits to the small-scale agriculture, are presented in Table 5.3.

Table 5.3 The main drivers of BBE from environmental, economic and social sustainability perspectives

	Drivers of BBE	*Reference*
Environmental sustainability	• Reducing exposure to high GHG emissions and high environmental risk	Andrews (2015)
	• Minimizing adverse environmental impacts	Ghisellini et al. (2016)
	• Reducing and/or complete elimination of wastes	Kulshreshtha et al. (2011)
	• Preserving the environment and maintaining the ecosystem services	Langeveld et al. (2010)
	• Increase the value of products by cost savings, e.g. through more sustainable resource use and making use of more agriculture waste resources and low-cost technologies that are operated and maintained locally.	Ghisellini et al. (2016)
	• Possibilities for new value creation, business growth and increase in margin and profits	Linder and Williander (2015)
Economic sustainability	• Potential for new business development, innovation and synergy opportunities (emergence of new investment opportunities for local entrepreneurs, e.g. trucking).	Kok et al. (2013)
	• Job creation and income generation by utilizing new resources and opening up new markets	EEA (2016)
	• Increased employment opportunities in rural areas, resulting in reduced outmigration of local people	Kulshreshtha et al. (2011)
	• *Access to new knowledge/information and innovation* (e.g. via social media): New technologies not only provide cleaner solutions for the future, but also help in avoiding problems caused by the current technologies	Ghisellini et al. (2016)
	• Information sharing platforms with many stakeholders enable better information transparency, thus help *adoption of BBE business models*	Ellen MacArthur Foundation (2013)
	• *Creating demands for perceived 'wastes' creates a trade of waste.* The potential trade-off between environmental and economic benefits created by policy-based BBE strategies uses economics as an implementation 'driver'	Ward et al. (2016)

	Drivers of BBE	*Reference*
Social sustainability	• *Increased crop diversification at farm level and FNS at household level.* SI promotes growing different types of crops in a given space over a period of time while providing multiple benefits for *better human nutrition and health* and promotes healthy and sustainable rural communities	Tesfai et al. (2019); Kulshreshtha et al. (2011)
	• Stricter regulations are examples of social drivers that increase pressure on agriculture to find alternatives from the existing LE business. Governmental support through laws and regulations have a clear enhancing role for BBE. Subsidies (e.g. the case of FISP in Malawi) and supportive taxation, such as recycling policies also lower agriculture risks in establishing new business around BBE	Tura et al. (2019)

The transition to a bio-based economy will take time and will not take place overnight (Metzea et al., 2017). It is a complex process involving fundamental changes to the production-consumption chain of the agriculture system (EEA, 2016) and must be technologically feasible, economically viable, and socially acceptable These changes have positive benefits on the environment, and on the economic and social aspects, which will result in more job creation, new opportunities, products and services (Figure 5.3).

The current linear economic model needs to be transformed to a bio-based economy model, but how to motivate farmers and other stakeholders? In fact, the BBE principles are not entirely new to the small-scale farmers of SSA. Because SSA farmers have already been applying some of

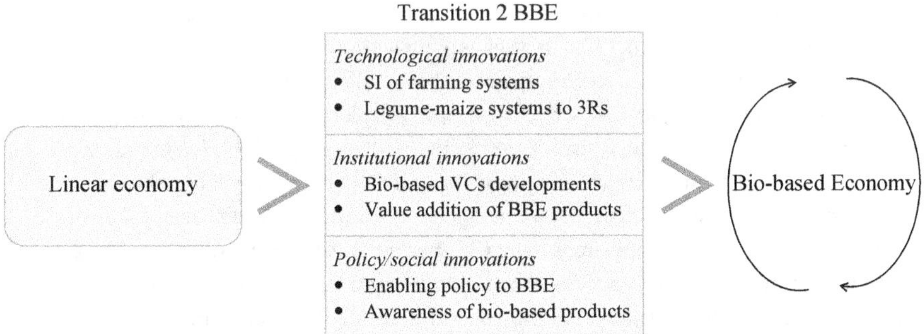

Figure 5.3 Framework conditions for transitioning an LE 2 BBE in the small scale agriculture sector.

the components of BBE on their farms, but perhaps not in an integrated or systematic way. The transition to a BBE needs a systemic change that encompasses innovations in technology (inclusive of VCs), institutions and policies (EC, 2017). Thus, it is necessary to raise awareness of the benefits, and create a framework condition under which BBE can flourish from technological, policy, institutional and/or cultural perspectives. The small-scale agriculture sector in Malawi (in this case, the maize-legume system or other similar systems) can contribute to a transition to BBE as follows.

Technological aspects of a transition

Innovative climate-resilient sustainable farming techniques: are knowledge-intensive and resource-demanding. They require adequate land, capital and labour, together with the appropriate information and a strong and enforceable property rights system. Technology options such as intercropping or crop rotation with legumes are easily adoptable in different agro-climatic and socio-economic conditions (see Box 5.1 as an example). To put it simply, any new technologies recommended should be of low risk and affordable by smallholders.

Institutional aspects of a transition

Institutional innovation should accompany the process of technology adoption along the VC. Programmes designed to overcome adoption barriers

> **Box 5.1 Adoption rate of different technologies in Malawi**
>
> A survey was conducted on the average adoption rate of the different innovative technologies by InnovAfrica project in Malawi. The average farm size was 0.89 hectares for 340 sampled households (HHs). Practices such as intercropping were adopted by 45 per cent; crop rotation by 20 per cent and cover cropping by 14 per cent. The higher adoption rate implies an efficient and intensive use of the land, compatible with the BBE principles on the use of resources. The adoption rate of planting pits, mulching, minimum tillage and tied ridges were between 2 and 7 per cent of the sampled HHs, which were lower than the legume-maize cropping practices. The survey also showed an interesting positive relation between access to inputs, credit and extension services and business opportunities (37 per cent of sampled farmers), and participation in group networks for dissemination of innovative practices (28 per cent of sampled farmers). Adoption rates of fertilizers and pesticides were 44 per cent and 27 per cent, respectively, while improved seeds were adopted by 75 per cent of the households interviewed.

should mainly focus on information and knowledge management; (public and private) institutions' effective extension services capable of *raising public awareness on bio-based products*; promoting wider dissemination of innovation among smallholders; and vertical integration and coordination along the VC (e.g. development of public-private partnerships).

Contract farming business models can be one option to provide the coordination needed in newly developing supply chains, to favour technology spill-overs, to guarantee market access and to improve employment opportunities in rural areas (Bennich et al., 2018; Branca et al., 2016). Contract farming on maize-legume VC may be a viable alternative farming model in Malawi, as it provides assured and reliable input service to farmers, and achieves the desired farm produce for the contracting firms and trading partners like NAFSAM (Kumwenda and Madola, 2005) with possible VC enhancement benefits. However, the risks associated with these models should not be transferred to smallholders. NASFAM (one of the InnovAfrica project Multi-Actor Platform members) has shown interest in doing business with the farmers in Dedza district, provided that the quantity and quality of agriculture produce are secured in a sustainable manner.

Policy/social innovations (including investments) aspects of a transition

Effective national policies should be modified, or new policies designed to strengthen the VCs, be it the maize-legume or other VCs. The policies should take into consideration both drivers affecting adoption of maize-legume interventions and the specific environmental and socio-economic barriers which affect the intervention types and the smallholders' needs and interests (Box 5.2).

Box 5.2 Policy Priority Area 4: the agricultural market in Malawi

Agricultural marketing systems in Malawi have suffered from several challenges, including deficient or missing infrastructure, policy and regulatory incoherence, and low private and public investments. These constraints have rendered Malawian agricultural value chains uncompetitive, nationally and regionally. The policy will therefore:

- Promote the growth and development of efficient and inclusive agricultural value chains that ensure competitive and fair pricing of agricultural commodities.
- Facilitate the creation of new structural markets, especially in legumes, oilseeds, sugar cane, coffee, horticulture, livestock, and fisheries products.

- Facilitate agricultural development and marketing corporation reforms to improve market efficiency and profitability to the benefit of farmers.
- Promote agricultural value addition and agro-processing.
- Support the growth and promote the competitiveness of agricultural commodity exchanges to enhance agricultural market efficiency.
- Strengthen and harmonize agricultural market information systems.
- Ensure that agricultural market and trade policies and regulations are transparent, evidence-based and enforced.
- Promote access to financial services and agricultural credit, particularly for producers and small and medium-scale agro-processors, including women, youth, and persons living with disabilities.
- Support improvements in quality standards and grading systems for all agricultural commodities.
- Promote regional and global exports of value-added agricultural commodities.

Source: MoAIW (Ministry of Agriculture, Irrigation and Water development) (2016)

In addition, policy incentives (e.g. long-term investment in farmland) that promote the implementation of the 3Rs principles of BBE are needed. This will help maintain soil cover, increase resource use efficiency, reduce GHG emissions, and increase production from the environmental aspects.

Conclusion

This chapter has reviewed and explored the potential of sustainable intensification (SI) of maize-legume cropping systems towards the transition to BBE. From the case study results of Dedza district, transforming small-scale agriculture to a sustainable model will be a challenge unless there are suitable technology inputs, adequate investments, enabling institutions and policies. There are numerous barriers along the bio-based VCs from the technology, marketing, cultural and policy/institutional perspectives. However, there are several drivers and opportunities to overcome the barriers to adoption of SI and then move towards BBE. Identifying the drivers and barriers to adoption of SI is the first step in the transition.

SI adoption within the maize-legume systems and integrated VCs development enhances resource use efficiency. Conditions for an integrated framework for transitioning to BBE under small-scale agriculture sector have been suggested in the chapter. The SI approach inclusive of VC developments is one potential option to transition to BBE in the small-scale agriculture context. By adopting the SI approach, the 3Rs principles of BBE could be effectively applied by adopting the following:

1 *At the production stage* by increasing on-farm production efficiency through crop diversification, reduced use of chemical inputs and substituting or supplementing them with farm manure, mulching with crop residues, to improve yields both in terms of quantity and quality with minimum impact on the environment. Farmers and stakeholders must be trained in the environmental benefits of SI systems.
2 *At the post-production stage* by:
   (a) strengthening the storage, marketing and processing facilities for smallholders in the local areas. For example, farmers should be able to store their produce and sell it when they get a good price. Credit vouchers could be provided against the stored produce that farmers could use to meet their household expenses.
   (b) organizing farmers into cooperatives, producer organizations, or contract farming, can help them to be better organized, and reduce their vulnerability to unstable markets. There are challenges associated with such models, which need to be addressed through stakeholder participation from the beginning. In addition, they should be provided with credit and extension services, mobile technology and ICT tools. Extension services should be improved, particularly for women and youth to adopt the SI systems.

However, to unravel the potentials of SI and promote sustainable pathways to move towards a BBE approach, further integrated research of the crop VCs such as maize-legume systems (including the environmental, social and economic drivers and barriers) is needed in the small-scale agriculture context of SSA.

Acknowledgement

We thank the University of Malawi team for their support in data collection.

References

AGRA (Alliance for a Green Revolution in Africa) (2013) *Africa Agriculture Status Report: Focus on Staple Crops*. Nairobi, Kenya: AGRA.

AGRA (Alliance for a Green Revolution in Africa) (2016) 'Africa agriculture status report, 2016: Progress towards Agricultural Transformation in Africa'. Available at: https://agra.org/aasr2016/public/assr.pdf

Andrews, D. (2015) 'The circular economy, design thinking and education for sustainability', *Local Economy*, vol. 30, no. 3, pp. 305–315.

Bennich, T., Belyazid, S., Kopainsky, B. and Diemer, A. (2018) 'The bio-based economy: Dynamics governing transition pathways in the Swedish forestry sector', *Sustainability* vol. 10, pp. 1–18.

Bisen, N. and Rahangdale, C.P. (2017) 'Crop residues management option for sustainable soil health in rice-wheat system: A review', *International Journal of Chemical Studies*, vol. 5, no. 4, pp. 1038–1042.

Branca, G., Cacchiarelli, L., Maltsoglou, I., Rincon, L., Sorrentino, A. and Valle, S. (2016) 'Profits versus jobs: Evaluating alternative biofuel value-chains in Tanzania', *Land Use Policy*, vol. 57, pp. 229–240.

Carus, M. and Dammer, L. (2018) 'The circular bioeconomy – concepts, opportunities and limitations, Hürth 2018–01'. Available at: www.bio-based.eu/nova-papers

CEC (Commission for Environmental Cooperation) (2014) *Burning Agricultural Waste: A Source of Dioxins*. Montreal, Canada: Commission for Environmental Cooperation, pp. 1–6.

Dorward, A. and Chirwa, E (2011) 'The Malawi Agricultural Input Subsidy Programme: 2005–2006 to 2008–2009', *International Journal of Agricultural Sustainability*, vol. 9, no. 1, pp. 232–247.

EC (European Commission) (2017) *Circular Economy Research and Innovation: Connecting Economic and Environmental Gains*, Brussels: EC.

EEA (European Environment Agency) (2016) 'Circular bioeconomy in Europe: Developing the knowledge base, Report No 2'. http://dx.doi.org/10.2800/51444

Ellen MacArthur Foundation (2013) 'Towards the circular economy: economic and business rationale for accelerated transition (1), Cowes'. Available at: www.thecirculareconomy.org

Ellis, F. and Manda, E. (2012) 'Seasonal food crises and policy responses: A narrative account of three food security crises in Malawi', *World Development*, vol. 40, no. 7, pp. 1407–1417.

FAO (Food and Agriculture Organization) (2011) *'Save and Grow': A Policymaker's Guide to the Sustainable Intensification of Smallholder Crop Production*. Rome: FAO.

FAO (Food and Agriculture Organization) (2017) *Regional Overview of Food Security and Nutrition in Africa 2016: The Challenges of Building Resilience to Shocks and Stresses*. Accra, Ghana: FAO.

Genovese, A., Adolf, A., Alejandro, F. and Koh, S.C.L. (2015) 'Sustainable supply chain management and the transition towards a circular economy: Evidence and some applications', *Omega*, vol. 66, pp. 344–357.

Ghisellini, P., Cialani, C. and Ulgiati, S. (2016) 'A review on circular economy: The expected transition to a balanced interplay of environmental and economic systems', *Journal of Cleaner Production*, vol. 114, pp. 11–32.

Gollin, D. (2014) 'Smallholder agriculture in Africa: An overview and implications for policy', IIED Working Paper, IIED, London.

ICRISAT (International Crops Research Institute for the Semi-Arid Tropics) (2009) 'Fertilizer micro dosing: Boosting production in unproductive lands'. Available at: www.icrisat.org/impacts/impact-stories/icrisat-is-fertilizer-microdosing.pdf

IPNI (International Plant Nutrition Institute) (2012) *4R Plant Nutrition: A Manual For Improving the Management of Plant Nutrition*, T. Bruuslema, et al. (eds) Norcross, GA: International Plant Nutrition Institute.

Jurgilevich, A., Birge, T., Kentala-Lehtonen, J., Korhonen-Kurki, K., Pietikäinen, J., Saikku, L. and Schösler, H. (2016) 'Transition towards circular bioeconomy in the good system', *Sustainability*, vol. 8, pp. 1–14.

Kok, L., Wurpel, G. and Ten Wolde, A. (2013) 'Unleashing the power of the circular economy'. Amsterdam: IMSA Amsterdam for Circle Economy April 2, 2013. Available at: http://mvonederland.nl/system/files/media/unleashing_the_power_of_the_circular_economy-circle_economy.pdf

Kulshreshtha, S., McConkey, B.G., Liu, T.T., Dyer, J.A., Vergé, X.P.C. and Desjardins, R.L. (2011) 'Bio-based economy: Sustainable use of agricultural resources, environmental impact of biofuels'. In M.A. Dos Santos Bernardes (ed.), *Environmental Impact of Biofuels*. London: IntechOpen Limited.

Kumwenda, I. and Madola M. (2005) 'The status of contract farming in Malawi', Food, Agriculture and Natural Resources Policy Analysis Network. Available at: www.fanrpan.org/archive/documents/d00691/Contract_Farming_Report.pdf

Langeveld, J.W.A., Dixon, J. and Jaworski, J.F. (2010) 'Development perspectives of the biobased economy: a review', *Crop Science*, vol. 50, no. 2, pp. 142–151.

Linder, M. and Williander, M. (2015) 'Circular business model innovation: Inherent uncertainties', *Business Strategy and the Environment*, vol. 26, no. 2, pp. 182–196.

Metzea, T., Schuitmakerb, J.T., Bitsch, L. and Broerseb, J. (2017) 'Breaking barriers for a bio-based economy: Interactive reflection on monitoring water quality', *Environmental Science and Policy*, vol. 74, pp. 1–7.

MoAIW (Ministry of Agriculture, Irrigation and Water Development) (2016) *National Agriculture Policy*. Lilongwe, Malawi: Malawi Government Ministry of Agriculture, Irrigation and Water Development, pp. 1–116.

Mungai, L.M., Snapp, S., Messina, J.P., Chikowo, R., Smith, A., Anders, E., Richardson, R.B., et al. (2016) 'Smallholder farms and the potential for sustainable intensification', *Frontiers in Plant Science*, vol. 7, p. 1720.

Pretty, J.N., Toulmin, C. and Williams, S. (2011) 'Sustainable intensification in African agriculture', *International Journal of Agricultural Sustainability*, vol. 9, no. 1, pp. 5–24.

Sasson, A. (2012) 'Food security for Africa: An urgent global challenge', *Agriculture & Food Security*, vol. 1, no. 2, pp. 1–16.

Schroeder, P., Anggraeni, K. and Weber, U. (2018) 'The relevance of circular bioeconomy practices to the sustainable development goals', *Journal of Industrial Ecology*, vol. 23, no. 1, pp. 77–95.

Tesfai, M., Njarui, D.M.G. and Ghimire, S.R. (2019) 'Sustainable intensifications of African agriculture through legume-based cropping and Brachiaria forage systems', *African Journal of Agricultural Research*, vol. 14, no. 27, pp. 1138–1148.

The Montpellier Panel (2013) 'Sustainable intensification: A new paradigm for African agriculture'. Available at: https://europa.eu/capacity4dev/hunger-foodsecurity-nutrition/document/sustainable-intensification-new-paradigm-african-agriculture

Thierfelder, C., Cheesman, S. and Rusinamhodzi, L. (2012) 'A comparative analysis of conservation agriculture systems: Benefits and challenges of rotations and intercropping in Zimbabwe', *Field Crops Research*, vol. 137, pp. 237–250.

Tura, N., Hanski, J. Ahola, T., Ståhle, M., Piiparinen, S. and Valkokari, P. (2019) 'Unlocking circular business: A framework of barriers and drivers', *Journal of Cleaner Production*, vol. 212, pp. 90–98.

United Nations (UN) (2015) 'Sustainable Development Goals'. Available at: https://sustainabledevelopment.un.org/content/documents/8521Informal%20Summary%20-%20UN%20Summit%20on%20Sustainable%20Development%202015.pdf

Vanlauwe, B., Hungria, M., Kanampiu, F. and Giller, K.E. (2019) 'The role of legumes in the sustainable intensification of African smallholder agriculture:

Lessons learnt and challenges for the future', *Agriculture, Ecosystems and Environment*, vol. 284, pp. 1–13.

Wang, B., Dong, F., Chen, M., Zhu, J., Tan, J, Fu, X., Wang, Y. et al. (2016) 'Advances in recycling and utilization of agricultural wastes in China: Based on environmental risk, crucial pathways, influencing factors, policy mechanism', *Procedia Environmental Sciences*, vol. 31, pp. 12–17.

Ward, S.M., Holden, N.M., White, E.P. and Oldfield, T.L. (2016) 'The 'circular economy' applied to the agriculture (livestock production) sector', discussion paper, presented at Workshop on the Sustainability of the EU's Livestock Production Systems, 14–15 September 2016, Brussels, Belgium.

World Bank (2014) 'World Development Indicator: Structure of output. Retrieved from data'. Available at: http://wdi.worldbank.org/table/4.2

6 Save and Grow

Sustainable intensification of crop production and innovative market links to build resilient rural economies in the Greater Mekong subregion

Jan Willem Ketelaar, Alma Linda M. Abubakar Avakat Phasouysaingam, Vornthalom Chanthavong, Ngo Tien Dung, Mayling Flores Rojas, Abha Mishra and Peter Sprang

Introduction

With the world's population expected to climb to 9.7 billion in 2050 from 7.7 billion today – meaning roughly 80 million more mouths to feed on an annual basis – the agriculture sector is under tremendous pressure to meet rising demand for food production. An estimated 820 million people still suffer from hunger worldwide (FAO, 2019a). Added pressures from climate change, extreme weather, shrinking and degraded agricultural lands, outbreaks of pests and diseases and the depletion of natural resources, make innovation and efficiency in support of enhancing productivity even more urgent (FAO, 2011, 2016). Yield growth rates of the major cereal crops, including rice produced by some 150 million smallholder farmers and consumed by some 800 million mostly poor people, have been declining steadily from 3.2 per cent in 1960 to 1.5 per cent in 2000 (FAO, 2009). If rice production is to increase and meet global demands in the future, smallholders will have to intensify production, albeit in a much more sustainable way compared to conventional – often mono-cropping-based – agriculture today.

One of the main concerns is the indiscriminate use of chemical inputs, both inorganic fertilizers and pesticides, particularly, in conventional rice production in Asia. The negative impact of chemical pesticides on human health, food safety and the environment, including reduced biodiversity and compromised ecosystem services, is well documented (Colborn et al., 1996; Kishi, 2005; Pretty and Hine, 2005). The Food and Agricultural Organization (FAO), the International Rice Research Institute (IRRI) and government-led national programmes have promoted the adoption of integrated pest management (IPM) and risk reduction strategies for pesticides in smallholder agriculture for decades, with some good results to show, albeit at still small-to-moderate level scales (Cochard et al., 2014; FAO, 2013; Gallagher et al., 2005). Despite long-standing scientific evidence that natural biological control can keep pests and diseases below economically important population levels, pesticide use remains the mainstay of pest control in Asian agriculture. It is further facilitated by outdated intensification policies, perverse subsidies on non-productive chemical inputs and weak regulations (as well as poor enforcement thereof), poorly performing and underfunded extension services. All this is boosted by aggressive and well-funded marketing campaigns by the chemical industry. Apart from environmental and human health concerns, this leads to reduced rice farming profitability, compromised national biological control, pest resistance development and the risk of aggravated pest outbreaks (Gallagher et al., 2005; Ketelaar et al., 2018).

Climate change, the most serious environmental challenge facing humanity today, is expected to have far-reaching impacts on the production of major crop commodities, including a projected reduction of 10 per cent in rice production by 2050 if climate-smart practices are not adopted by smallholder farmers. The rice sector itself is a major contributor to emissions of greenhouse gases (GHG) representing 10 per cent of all agriculture sector emissions worldwide. Mitigation efforts, including through wide-scale adoption of climate-smart practices, are therefore, urgently required (FAO, 2016; FOLU, 2019; UN, 2019).

In summary, rice farmers, particularly smallholders, are facing the major challenge of intensifying agricultural production to feed the world's growing population in the face of declining availability of labour, water and agricultural land, lower productivity and changing consumer patterns (FAO, 2011, 2016). In addition, this production must be sustainable with the minimum impact on the environment, including facilitating a substantial reduction in GHG emissions (FAO, 2010, 2019a). Some of these principles fall in line with the bioeconomy approach, which is based on the premise that future food production must be economically and environmentally sustainable.

In recent years, these challenges have led the FAO to intensively work with its member governments to promote *sustainable* intensification of agricultural production. The new paradigm emphasizes the need for a more efficient use of diminishing production resources and better management and use of agro-ecological processes, promoted by the FAO as the sustainable intensification of rice production (SIRP) under the banner of *Save and Grow* (S&G) (FAO, 2011). In 2016, the FAO published policy guidance for – and practical examples of – sustainable intensification of cereal crops, particularly, maize, rice and wheat (FAO, 2016). Other more Asia-specific socio-economic and cultural phenomena such as the 'greying' and 'feminization' of agriculture and vanishing rice cultures and heritage varieties have also led the FAO and its member country governments to examine and re-design policies and strategies to support a more *sustainable* rice production intensification.

This chapter, building on an earlier publication (Ketelaar et al., 2018), will summarize the recent work by the FAO and other partners on sustainable intensification of crop production. The chapter will present the results of project interventions, by the FAO and its associated partners, aimed at the adoption by farmers of good agricultural practices for SIRP. The chapter presents new research findings on farmers' adoption of good practices and results thereof that can be directly attributed to Save and Grow (S&G) farmer training interventions in the rice-based landscapes in Lao PDR, Vietnam and elsewhere in the Greater Mekong subregion. The case for increased policy support for – and public-private sector investments in – sustainable intensification of crop production and education programmes for smallholder farmers is outlined, aimed at the development of more inclusive 'Green Rice Landscapes' while contributing to achieving the multiple global UN Sustainable Development Goals (SDGs). Such investments are also vital for the development of robust and resilient rural economies that are based on key *circular bioeconomy* concepts, principles and good practices as outlined earlier in this book. This chapter's content also aligns well with and provides real-world examples of successful transformation towards greater sustainability in the agriculture sector (Box 6.1) as called for in recent landmark publications (FOLU, 2019; UN, 2019).

Box 6.1 The ten most critical transitions in food and land use systems necessary to achieve the Sustainable Development Goals

1 Promoting healthy diets.
2 Scaling productive and regenerative agriculture.
3 Protecting and restoring nature (and associated multiple ecosystem goods and services).
4 Securing a healthy and productive ocean.
5 Diversifying the protein supply.
6 Reducing food loss and waste.
7 Building local loops and linkages.
8 Harnessing the digital revolution.
9 Delivering stronger rural livelihoods.
10 Improving gender equality and accelerating demographic transition.

Source: Adapted from FOLU (2019).

The Regional Rice Initiative

In December 2012, the 145th Session of the FAO Council endorsed the formulation and implementation of a regional initiative for Asia to strengthen rice-based production systems as part of its new Strategic Objective 2: '*Make agriculture, forestry and fisheries more productive and sustainable*.' Subsequently, the FAO published a Regional Rice Strategy (RRS) and engaged in a strategic policy dialogue with the Asian and Pacific member countries, during the Asia and Pacific Regional Conferences convened by the FAO (FAO, 2014a). During the 2013–2017 period, the FAO also implemented a Regional Rice Initiative (RRI) working with selected member countries to apply the S&G concepts, principles and associated good practices in rice-based production landscapes in Asia. The RRI supported policy development and field implementation of the S&G approach in selected Asian countries, including Lao PDR and Vietnam. The S&G two-pronged approach and the intervention design, including the Farmers Field Schools' (FFS) capacity-building approach, were described and referenced in detail in an earlier publication by Ketelaar et al. (2018).

The RRI Save and Grow farmer education interventions were implemented through National Integrated Pest Management (IPM) programmes with a long and successful track record of capacity for large-scale FFS implementation. Farmers learned and field-tested good practices that are more productive, sustainable and efficient in resource use while promoting better *agro-ecology-based* intensification and diversification of rice-based farming systems. These good practices and more integrated system approaches included: rice-fish, rice-livestock and rice-vegetables/potato diversified production systems, IPM, agro-forestry, mechanization and promotion of climate-smart agriculture (CSA) practices such as the application of alternate wetting and drying (AWD) and use of rice straw instead of burning to mitigate emission of GHGs. The overarching S&G paradigm espouses agro-ecological approaches aimed at moving agricultural development away from a focus on single crop commodities and focal areas – e.g. improved seed, pest control and water management – to more holistic and agro-ecology-based solutions that promote diversification and reduced use of (chemical) inputs while integrating all components of the farming system (FAO, 2014b, 2015). This paradigm is very much aligned with the modern circular bioeconomy concept, principles and good practices. Thus, diversification on farms, in fact, helps the efficient use of resources and provides multiple income sources to the farmers. Therefore, it can be claimed that both the S&G and bioeconomy paradigms promote resource-efficient, sustainable food production.

Adoption results based on Save and Grow FFS in Lao PDR

During the 2015–2017 period, some 54 season-long Save and Grow-SIRP Farmers Field Schools (FFS) were implemented in Lao PDR within the

Box 6.2 A simple mechanization tool that saves time, money and effort: the drum seeder

In Lao PDR, specifically in Sayabouly Province, a labour-saving technology called a drum seeder was field-tested to support the sustainable intensification of rice production (SIRP) by the Laotian Government, smallholder farmers and supported by the Food and Agriculture Organization of the United Nations (FAO). A drum seeder is a manual technology used for drum seeding pre-germinated rice seeds, it is suitable for use by both men and women farmers. Based on the study results, drum seeding appears to be an attractive option compared to the traditional planting methods in the area, namely, manual transplanting and broadcasting. Using a drum seeder saves time and money in terms of labour and seeds, reduces the workload and the drudgery and increases labour productivity. Using a drum seeder also contributes to building resilience by smallholder farmers against climate change as they can re-plant their field faster if they lose their rice crops due to floods, droughts, etc.

Mr Noypan Phetsakone (a graduate of the Farmers Field School) explained, 'Before, ten people including myself and my wife would work for a whole day to transplant our rice field of 0.26 ha. Now with the drum seeder I can plant our field in 30 minutes.'

The drum seeder is an environmentally friendly technology as it does not require fossil fuel to operate. It is suitable for approaches such as rice-fish systems. Additionally, farmers can provide neighbouring farmers. with seeding services with the drum seeder. Those providing the service will generate an additional income while the farmers receiving the services will benefit from timely planting and reduced production cost in labour.

However, the drum seeder can increase the weeds when compared to broadcasting. The introduction of the drum seeder together with suitable weed management has proved successful in other countries in the region. This can be replicated when promoting the drum seeder.

To support its adoption in the rural communities, awareness of the benefits and limitations of the drum seeder is needed among local authorities, NGOs and other organizations working together with farmers in the field. More women should be included in the field demonstrations of the drum seeder, and women using other labour-saving technologies can increase their adoption in rural communities while contributing towards closing the gender gap.

Ms. Khamvean Senekhamleu said, 'It is good and easy to use the drum seeder, we save money on labour and time. Other women saw me using it and said that they would like to try too in their fields. With the days saved now I rest or go for other income activities.' The Senekhamleu family farm has been using the drum seeder on their rice field for the last three years.

Source: Adapted from FAO (2018).

context of the RRI (Figure 6.1). This involved action research and farmer education work in 17 districts spread over six provinces (Sayabouly (11), Xieng Khuang (10), Khammouane (1), Salavan (2), Savannakhet (17), Champassack (13)). Interventions took place, mostly (50 out of 54) during wet season rice production cycles. Each FFS met on a weekly basis on 17 occasions throughout the full production cycle. About 1,562 farmers (37 per cent female) participated in these FFS, where they learned about efficient management, growing healthy, well-yielding crops with fewer and more sustainable production inputs. Farmers explored making optimal use of multiple goods and services of paddy-based farming systems – conservation and management of aquatic biodiversity (including the introduction of cultured fish species) – in combination with improved agronomic practices such as wider plant spacing/reduced seeding rates, improved water management, and reduced use of chemical pesticides through the application of ecologically-sound IPM and natural biological control. Small-scale mechanization options, such as the use of drum seeders and harvesting machines were explored in all FFSs (see Figure 6.2). Further intervention details and results thereof as well as policy context were described in detail in the earlier published book chapter (Ketelaar et al., 2018) (see Box 6.2).

Figure 6.1 Save and Grow FFS in Lao PDR.

Source: ©FAO/Vornthalom Chanthavong.

Figure 6.2 Lao woman farmer in Sayabouly using the drum seeder.

Source: ©FAO/IPM the Lao People's Democratic Republic.

Two of these FFSs, involving some 50 rice farmers, were implemented in Phaxay district, Xieng Khouang province, in Lao PDR during the 2015 wet season. The farmers in Xieng Khouang province normally grow a local glutenous rice variety called 'Kao Kai Noi', suitable for production in the northern uplands. Compared to other varieties, Kao Kai Noi has a good yield and fetches higher farm gate prices. It is also a key ingredient in the brewing process of Lao Gold Beer, a value added for rice. In late 2018, with support from the FAO and the Lao government, this variety was formally registered with a Geographic Indication (GI), potentially setting up local rice farmers growing this variety in a registered GI association to benefit from higher farm gate prices and more stable market demands.

A results assessment study was carried out by the National University of Laos during the 2016 wet season rice production cycle, one year after the original FFS interventions in Phaxay district, Xieng Khuang. The aim of the study was to perform an initial assessment of the adoption of SIRP practices among FFS graduates in comparison to conventional (control) farmers in neighbouring villages. In terms of adoption of Save and Grow practices, FFS graduate farmers transplanted younger seedlings, used wider-spaced and substantially less seed inputs (74 kg/ha) compared to the non-FFS (87 kg/ha) and control (121 kg/ha) farmers, resulting in substantial savings on seed inputs. The assessment of the economic benefits of S&G adoption by farmers showed a 36 per cent higher gross margin than non-adopters. Details of the productivity gains and the economic benefits resulting from adoption of Save and Grow good practices are discussed in the earlier book chapters (Ketelaar et al., 2018, Phasouysaingam, 2017).

During the fourth quarter of 2017, the National University of Laos carried out a follow-up adoption study in Horkang and Phosy villages in Phaxay district of Xieng Khouang Province where FFS interventions had been undertaken in 2015. The study investigated the adoption and adaptation of S&G practices by Lao rice farmers in the first (2016) and second (2017) year after attending a wet season rice production FFS in Phaxay, Xieng Khouang, in 2015. The study used two sets of data that were collected in the same village and from the same respondents in 2016 and 2017 (Phasouysaingam, 2018). The measurement of adoption of Save and Grow practices by FFS graduate and non-FFS farmers was done according to a tailor-made scoring system as outlined in Table 6.1.

Table 6.1 Farmer adoption scores

Indicator	*Adoption score*
Y1 = Seed selection	- No selection applied = 1 - Good seed selection applied = 2
Y2 = Land allocated for S&G	- 0 = not allocated - 1 = Less than half of total paddy land area - 2 = more than half of total paddy land area
Y3 = Seedling age	- Less than 20 days after sowing (DAS) = 3 - Between 21 to 25 DAS = 2 - Older than 25 DAS = 1
Y4 = No. of seedlings/hill	- Equal and More than 5 = 1 - Between 4-1 = 2
Y5 = Planting space	- Less than 25 x 25 cm = 1 - More than 25 x 25 cm = 2
Y6 = Inorganic fertilizer	- Use more than 100 kg/ha = 1 - Use less than 100 kg = 2
Y7 = Manure/compost	- Apply manure = 1 - Not apply = 0

Source: Project field data.

Formula

$$\text{HHFAV} = \int_{j=1}^{3} b_j Y_j = b_1 Y_1 + b_2 Y_2 + b_3 Y_3 + b_4 Y_4 + b_5 Y_5 + b_6 Y_6$$

Note:
b = weight (b1 = 0.16, b2 = 0.14, b3 = 0.14, b4 = 0.14, b5 = 0.14, b6 = 0.14, b6 = 0.14)

Level of analysis

$$\text{Space} = \frac{\text{Max} - \text{Min}}{\text{No. level}} = \frac{1.86 - 0.72}{3} = 0.38$$

Note: Maximum score = 1.86, Minimum score = 0.72
Level 1 = Not-Adopted: score between 0.72 to 1.1
Level 2 = Partly adopted: score between 1.11 to 1.49
Level 3 = Adopted: score between 1.50 to 1.86

The adoption scores significantly increased for most types of S&G practices between 2016 and 2017 (Table 6.2). The S&G farmers adopted the good practices of younger seedlings for transplanting, wider spacing, lower number of seedlings/hill and more balanced and better applied mix of inorganic fertilizer and manure management, as recommended in the 2015 FFS. Only chemical fertilizer applications were not significantly different as farmers preferred to use chemical fertilizers according to their own experience of what had worked well in the past, although the FFS recommended using reduced rates of chemical fertilizer while applying more manure.

The results also showed that FFS graduate farmers were cautious adopters in the first year. The farmers adopted the new practices by splitting their land into S&G and conventional practices and evaluated performance

Table 6.2 Adoption scores of different S&G practices

S&G practices	*Min-Max score*	*Wet Season 2016*		*Wet Season 2017*	
		FFS graduate	*Non-FFS*	*FFS graduate*	*Non-FFS*
Seedling age	1–3	1.56 + 0.75	1.26 + 0.60	2.48 + 0.75**	2.04 + 0.87**
Planting space	1–2	1.37 + 0.49	1.38 + 0.50	1.70 + 0.47*	1.69 + 0.47*
No. seedling	1–2	1.56 + 0.51	1.31 + 0.47	1.89 + 0.32**	1.65 + 0.49**
Manure	0–1	0.85 + 0.36	0.46 + 0.51	1.85 + 0.36**	1.81 + 0.4**
Fertilizer	1–2	1.33 + 0.48	1.15 + 0.37	1.52 + 0.51^{ns}	1.15 + 0.37^{ns}

Source: Project field data.

Notes
T-Test comparing means for the practices between 2016 and 2017.* significantly different at 95%, ** significantly different at 99%, ns = not significant.

and results of application of different agronomic practices on rice yield and economic returns (see Table 6.3).

In the first year after FFS in 2016, FFS graduate farmers allocated 0.53 ha or 34 per cent of their land to S&G, while non-FFS allocated only 0.1 ha or 7 per cent. In the second year, the area of land under S&G increased to 0.73 ha or 46 per cent of their land for FFS graduate farmers, and 0.58 ha or 42 per cent for non-FFS farmers. The change between 2016 and 2017 was 38 per cent for FFS graduates, and 48 per cent for non-FFS farmers, which was highly significant at 99 per cent. The number of farmers who allocated land for S&G experiments increased significantly from 47 per cent in 2016 to 74 per cent of total respondents in 2017.

The number of adopting farmers increased from 8 per cent in 2016 to 60 per cent in 2017. In 2016, only 11 per cent of FFS graduates and 4 per cent of non-FFS farmers had adopted S&G practices. In 2017, these numbers increased to 74 per cent for FFS graduates and 46 per cent for non-FFS. The adoption for S&G practices referred to planting seedlings younger than 20 days of age, spacing at 25 x 25 cm or wider, planting only four or less seedlings per hill and applying more manure and less chemical fertilizer.

To summarize, the study investigated how farmers in Phaxay District adopted S&G practices after attending an FFS in 2015. The results showed that farmers were initially cautious and 70 per cent of the graduates only fully applied 11 per cent of S&G practices on 34 per cent of their farmland in 2016. By 2017, 85 per cent of these farmers had adopted 74 per cent of S&G practices on 43 per cent of their land. Other farmers in the same village who did not attend the FFS but saw the results also showed an interest in S&G practices. Only a few farmers (27 per cent) tried out S&G on a small portion (7 per cent) of their land. In 2017, however, 62 per cent of non-FFS farmers practised S&G on almost the same proportion of their farmland (42 per cent) as FFS graduates, but still lagged in the adoption of S&G practices (46 per cent). The adoption of S&G practices such as planting seedlings younger than 20 days of age, spacing the transplants at 25 x 25 cm or wider, and planting only four or less seedlings per hill was more prevalent among farmers. The adoption of applying more manure and less fertilizer, however, was only practised by 52 per cent of FFS graduates and 15 per cent of other farmers who saw the results in 2017.

The study did not assess the economic benefits of S&G adoption since it did not distinguish between the portion of farmland that was allocated to S&G and that which was not. However, the data showed that the application of S&G on less than half their farmland resulted in reduced seed and fertilizer costs while the expenses for manure increased. Overall, there was only a slight, non-significant cost reduction. However, those farmers who had adopted S&G practices had a 36 per cent higher gross margin than the non-adopters. There was a minor difference between partial and full adoption. If these farmers had applied S&G to

Table 6.3 Land allocation for Save and Grow practices between 2016 and 2017

Type of farmer	*Total land (ha)*	*Allocated land for S&G 2016 (ha)*	*No. of farmers*	*Allocated land for S&G 2017 (ha)*	*No. of farmers)*	*% of changing the land*	*T-Test*	*Percentage of change*
FFS graduate farmer (n = 27)	1.58 ± (0.67)	0.53 ± 0.76	19	0.73 ± 0.53	23	38	ns	FFS farmer 34
Non-FFS (n = 26)	1.39 ± 0.53	0.10 ± 0.23	7	0.58 ± 0.69	16	480	**	Non-FFS farmer 7
Total (n = 53)	1.44 ± 0.62	0.32 ± 0.60	25	0.66 ± 0.62	39	106	**	Total 22

Source: Project field data.

Notes

ns = not significantly different; ** significantly different by 99%; (...) = percentage of change.

their entire farmland, their benefits in terms of productivity and profits would likely have been even greater. The case study in Box 6.3 presents the benefits of applying S&G.

Box 6.3 FAO Model Farmer Award for Farmers Field School Save and Grow graduate

During the World Food Day celebrations at FAO in Bangkok in October 2018, Mrs Phonexay Thammavong of Horkang village, Phaxay district, Xieng Khuang Province, Lao PDR, was presented with the 2018 FAO Model Farmer Award (Figure 6.3). She received this Award in recognition of her outstanding achievement with regard to adoption of Save and Grow practices and helping other rice farmers in her community do the same.

After participation in a Farmers Field School during the 2015 wet season, Mrs. Phonexay adapted Save and Grow SIRP and was able to increase land productivity while making farming more profitable and sustainable through the more effective management and efficient use of farm inputs and natural resources. Through the application of Save and Grow techniques, her rice yield has increased by 100–150 per cent from 2.6 t/ha to an average of 4–5 t/ha, with a peak yield of 6.5 t/ha in 2018.

Mrs Phonexay did not use any chemicals in her rice paddy fields or in her vegetable garden. Instead, she applied the knowledge and skills acquired during her participation in the season-long Farmers Field School on Save and Grow SIRP to conserve natural enemies for natural biological control of crop pests, to make compost and use the manure collected from her livestock to fertilize the rice crop. She made botanical pesticides from available natural resources in the rice landscape to control pests and diseases in her rice paddy and vegetable crops.

Mrs Phonexay used 'Khao Kai Noi (KKN)' – a specialty, high quality glutinous rice variety – as the main crop in her own paddy fields. In addition, she has initiated agricultural diversification. Currently, she cultivates fish in the rice paddy (using her own bred fish), grows vegetables for family consumption all year round and sells excess vegetables at the local market, as well as keeping ducks, chickens, cows and horses. At her homestead, she also farms fish in a pond.

She is now able to grow organic rice which fetches a higher market price, allowing her to earn more income and make some saving. She has also helped other farmers in her community learn about sustainable rice production and integrated farming systems. Mrs Phonexay is now the head of an organic rice production group and has been involved in related activities, e.g. testing of good seed storage practices and the development of farmer association for Geographic Indication certification and marketing of specialty KKN rice. As a special honour, she was chosen to present her KKN

Figure 6.3 Lao FFS Save and Grow graduate, Mrs Phonexay, hands over KKN speciality rice to HRH Princess Maha Chakri Sirindhorn during a World Food Day event at the FAO in Bangkok on 19 October 2018.

Source credit: © FAO.

speciality rice to HRH Princess Maha Chakri Sirindhorn during a World Food Day event at FAO in Bangkok (Figure 6.3).

This case clearly demonstrates that farmers such as Mrs Phonexay can act as change agents and can significantly contribute to sustainable food production in the future.

The study has demonstrated impressively that FFS encouraged farmers to practise critical thinking and engage in experimentation. They did not blindly follow the Save and Grow recommendations, but first tried them out on a smaller scale and then expanded them quickly after they became more confident. Thus, it may take a few more years until the full potential of FFS education is realized by the entire village. Formal recognition by the government and the development partners of the pilot results from S&G FFS interventions can help raise awareness and facilitate scaling-up of sustainable intensification of crop production (see Box 6.3).

Adoption results based on Save and Grow FFS in Vietnam

The FAO supported the Vietnam National IPM Programme in partnership with the local civil society organization (CSO), while the Centre of

Initiatives on Community Empowerment and Rural Development (ICERD) carried out training on agro-aquatic (including rice-fish) and agrobiodiversity and pesticide impact assessment in selected communities (Towards a Non-Toxic Southeast Asia, 2016). Initial surveys carried out by ICERD showed that farmers in selected sites in Bac Giang and Quang Binh provinces in northern Vietnam practise rearing fish in ponds close to rice fields. Some farmers reportedly released fish from the ponds into the fields but only after harvesting the rice. Pesticide use – about 5–7 applications per crop, not including herbicides – made raising fish in standing rice crops impossible. As part of the process of community education on pesticide risk reduction, the National IPM Programme assisted the formulation of Community Action Plans and trained farmers in FFS in 2014 to reduce pesticide risk, including the use of biological control options for pest management. Among other agreements, various stakeholders in the commune – including the local government, people's organizations (Farmers' Union, Women's Union, Youth Union, etc.) – and farmers agreed to reduce the use of pesticides for improved rice ecosystem health and to integrate rice-fish-aquatic biodiversity in the production system.

A total of 105 FFS farmers (of which 55 were women) from Bac Giang and Quang Binh applied their knowledge and skills on improved rice intensification practices to an area of 34 hectares. Rice fields adjacent to each other were selected for S&G study plots. By following the good principles and practices of S&G, rice farmers applied efficient crop management, growing healthy, well-yielding crops with fewer and more sustainable pest management practices.

Farmers explored making optimal use of multiple goods and services of paddy-based farming systems – including conservation and management of aquatic biodiversity (both captured and cultured fish species) – in combination with improved agronomic practices, such as wider plant spacing/reduced seeding rates, improved water management, and reduced chemical pesticides through the application of ecologically sound IPM. This included the use of biological control agents such as entomopathogenic fungi, including *Metarhizium anisopliae*, as an alternative to chemicals in the management of the brown planthopper (*Nilaparvata lugens*) and natural biological control provided by ecosystem services available free of charge in healthy rice landscapes.

Table 6.4 summarizes the differences between yields and benefits farmers obtained if the farmers grew only rice and when aquatic biodiversity was integrated in the production system.

The average gross income from integrated rice-fish-aquatic production was US$7,751 compared to US$1,892 obtained from producing only rice. Using integrated rice-fish-aquatic biodiversity production practices resulted in an average gross income ranging from 211–551 per cent compared with producing rice only. Farmers' experience and the aquatic biodiversity

Table 6.4 Differences between yields and benefits in rice-only and integrated agro-aquatic biodiversity production systems

Parameters	*Total Bac Giang*	*Total Quang Binh*	*Average*
A. Rice yields (kg/ha)	6,120	5,417	5,769
B. Gross income from rice production only (US$/ha)	2,215	1,569	1,892
C. Yields of fish and other aquatic organisms (kg/ha)	7,913	1,860	4,886
D. Gross income from fish and other aquatic organisms (US$/ha)	9,981	1,738	5,860
E. Gross income from rice, fish and other aquatic organisms [B + D] (US$/ha)	12,196	3,307	7,751
F. Input costs (US$/ha)	4,547	1,402	2,975
G. Profits [E – F] (US$/ha)	7,649	1,905	2,975
H. Difference in gross income between rice production only and integrated rice-fish-aquatic biodiversity production (% increased)	551	211	381

Notes
Exchange rate: US$1: VND21,405; Source: Project field data.

species and numbers – especially fish – accounted for the big difference in benefits. Informed management decision-making based on agro-ecosystem analysis, the use of ecosystem services such as natural pest control to avoid unnecessary use of chemical pesticides and the use of biological control agents also contributed to effective and sustainable pest management.

The Save and Grow FFS in Vietnam provided real opportunities for farmers and other community stakeholders to field-test rice intensification practices, making optimal use of the multiple goods and services available from and produced by paddy-based farming systems. Diversification of rice-based farming systems, including the inclusion of zero-tillage potato production, allowed farmers to raise their productivity, reduce labour inputs and gain higher profits while mitigating emissions of GHGs (Dung et al., 2018; FAO, 2015).

Adoption of system of rice intensification in the Greater Mekong subregion

Another relevant example is the European Union-funded project, called 'Sustaining and Enhancing the Momentum for Innovation and Learning around the System of Rice Intensification (SRI) in the Lower Mekong River Basin (LMB)', implemented by the Asian Institute of Technology (AIT) during the period 2013–2018. The FAO and associated government

National IPM-FFS Programmes helped implement this regional project in the Greater Mekong subregion countries of Cambodia, Lao PDR and Vietnam. The main objective of the project was to engage rainfed rice farmers in field testing of system of rice intensification (SRI) practices, a recognized Save and Grow-based suite of best practices for sustainable intensification of rice production (FAO, 2016). The project also facilitated building strong farmer networks at the community level in rainfed production areas.

In contrast to traditional methods of rice cultivation, SRI techniques require less water, fewer seeds, less manure, and less labour and promise higher yield and economic returns. The SRI principles were applied in both manual and mechanized paddy cultivation, depending on the location-specific requirements and farmers' choice. In several countries, such in Thailand and Cambodia, the SRI adaptation involved mechanization in smallholder farmers' fields where direct seeding machines and drum seeders were introduced to save the labour and seed costs without compromising yields. In direct seeding SRI, fewer seeds per hole were applied, maintaining 20x20 cm spacing. To deal with the variabilities of the weather, such as early droughts and late flooding which is quite common in the region, farmers also adjusted paddy planting time to harness the benefits of SRI principles. Water management, as per SRI principles and practices, including the use of AWD, was challenging since most of the farmers were cultivating rice in rainfed environments. However, in some irrigated systems, such as in Uttaradit in Thailand and Bac Giang in Vietnam, farmers reduced the frequency of irrigation and appreciated SRI water management principles, including the application of AWD, for better effects on plant growth. They reported better root growth, better tillage and higher yields under alternate wet and dry soil environment compared to the continuously flooded soil condition.

The action involved more than 15,000 farmers (> 50 per cent women) in the farmer-led field trials located in 33 districts of 11 provinces of the 4 countries (Cambodia, Laos, Thailand and Vietnam). The project results showed that in comparison with the pre-project baseline, SRI practices helped to improve livelihoods and the environment across the LMB region in numerous ways, for example:

- The average rice yield increased by 52 per cent, and net economic returns increased by 70 per cent.
- Labour productivity increased by 64 per cent, water productivity by 59 per cent, and fertilizer use-efficiency by 75 per cent.
- The total energy input required for farming operations decreased by 34 per cent, along with significant reductions in per-hectare net emission of greenhouse gases, respectively by 14 per cent with irrigated rice production, and by 17 per cent in rainfed cropping (Mishra, 2019a).

The SRI-LMB project facilitated the development of informal farmers' groups in 11 provinces across all four countries. These groups, if strengthened further, can provide a basis for developing farmer organizations that can accelerate green growth and sustainable rice intensification and crop diversification in support of building more circular bio-economies in the LMB countries.

Discussion

The results of the Save and Grow SIRP FFS interventions have clearly shown the enormous potential of rice farmers to make more efficient use of production inputs while raising land productivity and their incomes. The evidence on adoption of S&G practices by Farmers Field School graduates in Lao PDR and Vietnam and elsewhere in the Greater Mekong subregion as presented in this and earlier book chapters (Dung et al., 2018; Ketelaar et al., 2018) clearly illustrate that farmers can intensify production and make rice farming more profitable and climate-smart while at the same time becoming better custodians of the multiple goods and ecosystem services provided by healthy rice-based farming systems and landscapes. The FFS interventions have also made a substantial contribution towards the development and publication of evidence-based case studies in support of policy advice and communication for development endeavours by the FAO and other partners in support of food security and nutrition (FAO and CAAS, 2018). While the RRI has focused on sustainable intensification of rice production in the *irrigated* lowlands, the SRI-LMB project has clearly shown that also farmers in *rainfed* production areas can raise their productivity and intensify rice production in a sustainable manner with good productivity and profit gains evidenced.

The positive results and impact of the Save and Grow capacity building interventions can be attributed to the adult education approach used in the Farmers Field Schools. This farmer education approach is ideally suitable and effective for smallholder farmers to learn about knowledge-intensive innovations and sustainable management of natural resources (Swanson and Rajalahti, 2010). Farmers Field Schools facilitate ecosystem literacy training which is essential for smallholder farmers to acquire the required agro-ecological knowledge and skills needed for sustainable rice production intensification (Ketelaar and Abubakar, 2012). Farmers Field Schools, with their roots in Asian rice paddy fields and developed by the FAO 30 years ago, have evolved to address a wide range of topics and are now implemented in over 90 countries worldwide (FAO, 2019b).

Key development areas explored within the context of FFS and the associated action research capacity building interventions addressed the need for farmers to better understand and apply soil health concepts, principles and good practices, such as AWD. The aerobic soil conditions facilitated by AWD enhances the diversity of favourable microbial activity, supports nutrients recycling and allows better retention of water and nutrients. This in turn encourages favourable soil-root-plant interactions for better plant and soil productivity (Mishra, 2019b; Mishra et al., 2006). There is also substantial potential for better integration of green manure crops, including legumes, into cropping calendars and adoption of improved practices for straw management, including use of straw as mulch, in support of development of more diversified, less environmentally polluting and more profitable rice-based farming systems (Dung et al., 2018).

Promotion of climate-smart practices, including for mitigation and adaptation, featured prominently in the RRI and SRI-LMB capacity-building interventions. Adoption of alternate wetting and drying has the potential to reduce at least 10 per cent of regular methane emissions compared to keeping rice fields continuously flooded (Mishra, 2019b). A book chapter published earlier on zero-tillage potato production in rice-based landscapes in Vietnam (Dung et al., 2018) provided examples of highly successful and productive interventions aimed at the mitigation of GHG emissions (use of straw for mulching instead of burning), potentially making a substantial contribution towards the Vietnam government's GHG emissions reduction targets.

The use of mechanization in modern rice production in Asia is increasing and its importance cannot be underestimated. Mechanization in both rainfed and irrigated rice production in the Greater Mekong subregion can provide much-needed support to enhance input use efficiencies and reduce manual labour inputs in land preparation, crop establishment, weeding, harvesting and post-harvest/processing operations. Small-scale mechanization should reduce the manual labour inputs by women who have the double burden of working in the rice fields as well as caring for their households. Mechanization options should be environmentally competitive, climate-smart, economically viable and affordable. This should also target youth specifically to make agriculture more attractive for employment and entrepreneurship. The Save and Grow SIRP FFS interventions promoted the testing and application of simple mechanization options to reduce labour and drudgery from manual transplanting, weeding and harvesting.

Full mechanization along the entire value chain includes enhancing access to mechanization services, improving access to quality, reliable supply and affordable inputs, such as water, seed and fertilizer, and more efficient use thereof, including the use of modern technologies, such as robotics and drones,[1] and delivering efficient water resources management systems, including irrigation. Special attention is needed to reduce harvest and post-harvest losses, through improved threshing, drying, and storing, adding value through milling, processing and packaging, and improving market access through transportation.

For SIRP to be scaled up beyond pilot intervention areas, the Asian countries should provide an enabling policy environment while abolishing perverse subsidies on chemical production inputs that continue to facilitate the inefficient use of resources. Better regulations governing chemical input and their enforcement thereof, including the ban on and restriction of the most toxic chemical pesticides, have a proven track record of reducing health risks from pesticide to humans.[2] Equally important, jointly with resource partners, the government should promote investments in relevant capacity-building interventions, preferably based on FFS-based farmer education and ecosystem literacy training. The increased international recognition in recent years of the substantial contribution of the rice production sector in terms of emissions of GHG and the potential for reductions through the adoption of climate-smart practices should facilitate a more enabling funding environment (for example, the Global Environment Fund (GEF) and Green Climate Fund (GCF) financing) for such endeavours.

Clearly, relevant public-private sector partnerships can and must perform a greater role in transforming the rice sector towards greater sustainability. The FAO is now engaged in a Sustainable Rice Landscapes Initiative which represents a unique consortium of public, private and civil society partners, bringing together technological, ecological, policy and market-led approaches to the challenges of rice sustainability. Resource mobilization efforts by the FAO and other relevant UN agencies and research and development partners, including through GEF and GCF funding modalities, are now underway.

Rice value chains can benefit from the adoption of sustainable practices such as IPM, as defined in the world's first voluntary sustainability standard for rice developed by the Sustainable Rice Platform (SRP). This public-private sector partnership (see Box 6.4), co-convened by the UN Environment and IRRI with over 100 member organizations and institutions to date, aims to promote sustainable rice cultivation in the international rice trade. Whereas, the current share of *internationally* traded rice is less than 10 per cent, with the remaining 90 per cent of rice

Box 6.4 The Sustainable Rice Platform

The SRP is a global multi-stakeholder alliance launched in 2011 by UN Environment and the International Rice Research Institute (IRRI) with strong private sector participation. As of 2019, the alliance includes over 100 members, ranging from financial institutions to NGOs. SRP's goal is to minimize the environmental impacts of rice production and consumption while enhancing smallholder incomes and contributing to food security.

The wide-scale adoption of resource-use efficiency and climate-smart agriculture in rice systems is promoted by SRP both on farms and through value chains. SRP pursues voluntary market transformation initiatives by providing standards, indicators, an incentive-driven assurance scheme and outreach mechanisms supported by the international rice trade.

The SRP Standard on Sustainable Rice Cultivation, first launched in 2015, has been the most widely used SRP tool. Version 2.0 was released in 2019 after a two-year public consultation process and comprises of 41 requirements structured under eight themes

1. Farm Management
2. Pre-planting
3. Water Use
4. Nutrient Management
5. Integrated Pest Management
6. Harvest and Post-harvest
7. Health and Safety
8. Labour Rights.

The requirements restrict, for example, the use of agrochemicals ('no broad-spectrum insecticide application within the first 40 days after planting') and the burning of straw, as well as encouraging aerobic decomposition of organic material to avoid greenhouse gas emissions. Apart from higher farmer income through efficient input management, the Standard prescribes quality seed material and food safety aspects, that position producers as viable business partners in the rice value chain and international markets with agrochemical residue limits.

produced sold in *domestic* markets, the potential spill-over of the adoption of international sustainability standards into the domestic trade should not be underestimated. If standard adoption is genuinely promoted, including by private sector stakeholders, to reduce agrochemical input use, then the rice sector stands a fair chance of moving forward towards greater sustainability in support of building better rural livelihoods as part of more circular bio-economies at scale in Asian rice landscapes.

Conclusion

In summary, the key lessons learned, recommendations and policy advice based on from recent initiatives by the FAO and other development partners aimed at the promotion of sustainable intensification of rice production landscapes are presented below:

- Sustainable rice production intensification is feasible, allowing farmers to both *Save* (on input costs while protecting vital ecosystem goods and services) *and Grow* (raise land productivity and crop yields).
- Improved management practices, use of more productive and affordable production inputs, mechanization and associated reduced labour inputs and responsible management of agro-biodiversity bring significant increases in yields, reductions in costs and GHG emissions and

higher rural incomes. The GHG reduction potential of S&G methods will require further research to establish the claims. This could provide incentives for both older but especially younger generations to gain a renewed interest in rice-based production systems and securing livelihoods in rural areas.

- Recognizing the knowledge-intensive nature of S&G, smallholder farmers trained in season-long FFS have been shown how to master the agro-ecology and sustainability concepts and management skills required for sustainable use of natural resources and how to apply and adapt sustainable practices in crop management with good results.
- Investments in farmer education and ecosystem literacy training, preferably through FFS, are vital for large-scale adoption and scaling up of the good results of S&G.
- Development of innovative market links, including through the protection and potentially more lucrative sales of specialty rice varieties through Geographic Indication certification schemes, can help motivate farmers to sustain the application of these practices while raising rural incomes.
- The private sector can and should play a major role in facilitating the scaling-up of the adoption of resource-use efficient and climate-smart practices in rice production. In particular, the private sector and major rice buyers should commit to 100 per cent sourcing of SRP standard-based rice. Pro-active action by the private sector can help steer the entire rice sector towards greater sustainability.

Notes

1 The global market for agricultural drones and robots is projected to grow from $2.5 billion in 2018 to $23 billion in 2028, according to a report from market intelligence firm BIS Research: https://bisresearch.com/industry-report/agriculture-drones-robots-market.html

2 For example, Sri Lanka banned a series of the most toxic pesticides over the course of two decades and has seen a substantial reduction (70 per cent) in the number of deaths resulting from intentional suicides (Lancet Global Health, 2017, cited in BBC, 2019).

References

BBC News (2019) 'Which countries have banned pesticides used in suicide?', 15 September. Available at: www.bbc.co.uk/news/world-asia-49662340

Cochard, R., Maneepitak, S. and Kumar, P. (2014) 'Aquatic faunal abundance and diversity in relation to synthetic and natural pesticide applications in rice fields of Central Thailand', International Journal of Biodiversity Science, Ecosystem Services & Management, vol. 10, no. 2, pp. 157–173.

Colborn, T., Myers, J.P. and Dumanoski, D. (1996) *Our Stolen Future: How Man-Made Chemicals Are Threatening Our Fertility, Intelligence and Survival.* London: Little Brown and Company.

Dung, N.T., Nagothu, U.S., Morales-Abubakar, A.L., Ketelaar. J.W. and Tesfai, M. (2018) 'Innovative practices in potato production for food and nutrition security', in U.S. Nagothu (ed.) *Agricultural Development and Sustainable Intensification: Technology and Policy Challenges in the Face of Climate Change.* London: Earthscan, pp. 111–135.

FAO (Food and Agriculture Organization) (2009) *How to Feed the World in 2050*, Rome: FAO.

FAO (Food and Agriculture Organization) (2010) 'Sustainable crop production intensification through an ecosystem approach and an enabling environment: capturing efficiency through ecosystem services and management'. Report of the 22nd session of the Committee on Agriculture. Rome: FAO.

FAO (Food and Agriculture Organization) (2011) 'Save and Grow: A policymaker's guide to the sustainable intensification of smallholder crop production.' Available at: www.fao.org/3/a-i2215e.pdf

FAO (Food and Agriculture Organization) (2013) 'Empowering farmers to reduce pesticide risks'. Available at: www.vegetableipmasia.org/docs/Empowering%20 Farmers%20To%20Reduce%20Pesticide%20Risks%2028Oct.pdf

FAO (Food and Agriculture Organization) (2014a) 'A regional rice strategy for sustainable food security in Asia and the Pacific'. Available at: www.fao.org/3/i3643e/i3643e.pdf

FAO (Food and Agriculture Organization) (2014b) 'The multiple goods and services of Asian rice production systems'. Rome: FAO.

FAO (Food and Agriculture Organization) (2015) 'Stories from the field: Vietnam, Save and Grow'. Available at: http://vegetableipmasia.org/docs/Case%20Studies/Save%20and%20grow%20Final.pdf

FAO (Food and Agriculture Organization) (2016) *Save and Grow in Practice: Maize, Rice, Wheat: A Guide to Sustainable Cereal Production.* Rome: FAO.

FAO (Food and Agriculture Organization) (2018) 'Gender-sensitive labour saving technology: Drum seeder saving time, effort and money'. Available at: www.fao.org/3/i9464en/i9464en.pdf

FAO (Food and Agriculture Organization) (2019a) 'The state of food security and nutrition in the world: Safeguarding against economic slowdowns and downturns'. Available at: www.fao.org/3/ca5162en/ca5162en.pdf

FAO (Food and Agriculture Organization) (2019b) 'Farmers taking the lead: thirty years of Farmer Field Schools'. Available at: www.fao.org/3/ca5131en/ca5131en.pdf

FAO and CAAS (Food and Agriculture Organization and the Chinese Academy of Agricultural Sciences) (2018) 'Agroecology for food security and nutrition'. Proceedings of the Agroecology Workshop in China, Kunming, Yunnan, 29–31 August 2016. Rome: FAO.

FOLU (Food and Land Use Coalition) (2019) 'Growing better: Ten critical transitions to transform food and land use'. Global Consultation Report. Available at: www.foodandlandusecoalition.org/wp-content/uploads/2019/09/FOLU-Growing Better-GlobalReport.pdf

Gallagher, K., Ooi, P., Mew, T., Borromeo, E., Kenmore, P. and Ketelaar, J.W. (2005) 'Ecological basis for low-toxicity Integrated Pest Management (IPM) in rice and vegetables', in J. Pretty (ed.) *The Pesticide Detox*. London: Earthscan, pp. 116–146.

Ketelaar, J.W. and Abubakar, A.L. (2012) 'Sustainable intensification of agricultural production: The essential role of ecosystem-literacy education for smallholder farmers in Asia', in K. Mijung and C.H. Diong (eds) *Biology Education for Social and Sustainable Development*. Amsterdam: Sense Publishers, pp. 173–182.

Ketelaar, J.W., Morales-Abubakar, A.L., Van Du, P., Widyastama, C., Phasouysaingam, A., Binamira, J. and Dung, N.T. (2018) 'Save and Grow: Translating policy advice into field action for sustainable intensification of rice production', in U.S. Nagothu (ed.) *Agricultural Development and Sustainable Intensification: Technology and Policy Challenges in the Face of Climate Change*. London: Earthscan. pp. 23–51.

Kishi, M. (2005) 'The health impacts of pesticides: What do we know?', in J. Pretty (ed.) *The Pesticide Detox*. London: Earthscan, pp. 23–38.

Mishra A. (2019a) 'Boosting yields, raising incomes, and offering climate-smart options: The system of rice intensification paves the way for farmers to become more successful "agripreneurs"'. Available at: www.sri-lmb.ait.asia/downloads/SRI-LMB%20Final%20Report2019.pdf

Mishra, A. (2019b) 'Morphological and physiological root plasticity and its relationships with shoot growth of rice', in O. Takuji (ed.) *Root Biology: Growth, Physiology, and Functions*. doi: http://dx.doi.org/10.5772/intechopen.87099.

Mishra, A., Whitten, M., Ketelaar, J.W. and Salokhe, V.M. (2006) 'The System of Rice Intensification (SRI): A challenge for science, and an opportunity for farmer empowerment towards sustainable agriculture', *International Journal of Agricultural Sustainability*, vol. 4, no. 3, pp. 193–212.

Phasouysaingam, A. (2017) 'Case study on FFS graduate farmer adoption of Save and Grow (S&G)-based sustainable intensification of crop production', unpublished consultancy report submitted to FAO. National University of Laos, Vientiane.

Phasouysaingam, A. (2018) 'Farmer adoption study report on FFS Save and Grow (S&G)-based sustainable intensification of crop production', unpublished, consultancy report submitted to FAO. National University of Laos, Vientiane.

Pretty, J. and Hine R. (2005) 'Pesticide use and the environment', in J. Pretty (ed.) *The Pesticide Detox*. London: Earthscan, pp. 1–22.

Swanson, B.E. and Rajalahti, R. (2010) 'Strengthening agricultural extension and advisory systems: Procedures for assessing, transforming, and evaluating extension systems'. Agriculture and Rural Development Discussion Paper no. 45, Washington, DC: The International Bank for Reconstruction and Development and the World Bank.

Towards a Non-toxic Southeast Asia (2016) 'Agro-biodiversity and rice-fish farming'. Available at: www.sida.se/globalassets/sida/sve/har-arbetar-vi/stories-from-the-field-2016.pdf

United Nations (UN) (2019) 'Global Sustainable Development Report 2019: The future is now – science for achieving sustainable development'. Available at: https://sustainabledevelopment.un.org/globalsdreport/2019

7 Protected cultivation of vegetables in Sub-Saharan Africa

Scope and impacts

Hugo Despretz, Thibault Nordey and Armel Clement Goudjo Mensah

Introduction

Agriculture is connected to several societal challenges, such as food security, fossil-fuel dependence, natural resource scarcity, and climate change. As agriculture relies on the use of renewable biomass and bioprocesses (plants, animals, and micro-organisms), it is one of the sectors where bioeconomy can be promoted, if it is organized in a sustainable manner. However, there are different ways of production that impact the environment in different ways. After the Second World War, the agricultural economy in the West underwent a deep transformation towards production intensification. Technological advances, such as the introduction of machinery powered by petrol and electricity and the widespread use of pesticides and chemical fertilizers, dramatically increased crop productivity but had detrimental impacts on the health of both the producers and the consumers and also on the environment (including soil erosion, water pollution, decline in rural biodiversity and other related effects). A similar trend followed in the East, in the 1960s and the 1970s, because of the Green Revolution where agricultural expansion happened at the cost of the environment. The focus at that time was only on a few grain crops, including wheat, rice and maize. It was not only pollution from the agricultural intensification but also deforestation to bring new areas under cultivation that led to unsustainable pathways of production and loss of biodiversity.

A global awareness in society and the demand to orient agricultural production systems towards more sustainable ways intensified by the early 1980s. This has led to the development of agro-ecological and other similar systems where emphasis was put on using technology in conjunction with natural, social and human conditions (Malézieux et al., 2009). At the same time, attempts were made to produce not only cereals, but also to promote crop diversification for the sake of healthy diets. This gradually promoted the large-scale cultivation of fruits and vegetables as they are an important source of nutrients in balanced diets which are necessary for human health. However, the indiscriminate use of pesticides and mineral fertilizers to increase vegetable and fruit production negatively affected the environment and human health. The high profitability of vegetable crops and their high sensitivity to climatic conditions and pests motivated the intensification of cropping systems. Among the technically most advanced production systems, protected cultivation (PC) of crops emerged as a promising yet improvable approach to achieve agricultural sustainability as part of a global bioeconomy.

The PC of crops in many parts of the world contributed to an increase in farm income, job creation and rural development as the market demand for vegetables, fruits and flowers increased. Thus, PC contributed to the bio-based economy directly or indirectly in several countries, including those in Sub-Saharan Africa (SSA). However, the dependence on technology and the initial investments prevented smallholders, especially in the developing countries, from taking advantage of the PC system, until alternative low-cost methods were made available. Sustainability of PC models largely depends on the type of inputs and the technology in use. In the first section of this chapter an overview of the PC development is presented. It then summarizes two case studies from SSA and their implications. The constraints for scaling up protected cultivation of vegetables (PCV) in SSA are then discussed. Finally, key recommendations are provided to show how sustainable PCV can promote bioeconomy in the region.

Overview of protected cultivation systems worldwide

The technological evolution of protected cropping systems

Protected cultivation of plants, meaning the use of covers directly on the soil and/or above crops, such as small or high tunnels (Figure 7.1), is an age-old method of farming that today has gained tremendous momentum worldwide. The intention is to create conducive conditions for plant

Figure 7.1 Tomatoes grown via hydroponics, Nantes region, France.
Source: photo by J.M. Mahé.

production by modifying the natural environment, using devices, technologies, and structures that can influence the climatic, hydrological and biotic parameters around the plants. Crops are being artificially protected in different ways depending on the agro-climatic conditions: in temperate or cold conditions, the aim is mostly to retain heat and light around the crop, whereas, in tropical or arid conditions, the objective is to protect against pests and excess sun radiation or rainfall. Therefore, various technologies are integrated to address the challenges that growers face in open air conditions, more so in extreme weather conditions.

The late fifteenth century witnessed the initial rudimentary technologies and techniques in Europe and East Asia, aimed first at protecting the plants from the cold and thereby extending the production periods. From then on, simple wooden squares covered with glass panes, glass bells or oiled paper have evolved into erected structures sometimes heated by stoves or heating flues. The design of the primitive greenhouses has slowly evolved to optimize the capture of light and heat to allow the growth of various crops, from low ones (lettuce, radishes) to climbing vines (cucumbers, grapes). Consequently, greenhouse horticulture developed to provide different options to the farmers. For example, provisions were made to reduce excess energy (heat) within the structure that would be discharged by ventilation and similarly, any deficits or surplus in the temperature would be compensated for by heating or cooling (Graamans et al., 2018).

With the invention and proliferation of synthetic polymers in the twentieth century, the agricultural sector worldwide witnessed a new kind of agriculture technology development termed plasticulture. Plastic has made its way into irrigation systems, mulches, potting, packaging, greenhouse covering materials, and so forth, optimizing agricultural production systems (Lamont, 2005). Cheaper than glass, easier to handle, lighter, more flexible (permitting new greenhouse designs) and with similar optical properties, plastic structures have contributed to greenhouse farming spreading and diversifying its technologies: from nets with various mesh sizes preventing the penetration of specific insects and allowing natural ventilation to spectral selective or light-diffusing films, growers can now choose from a large array of plastic covers to appropriately protect their crops from the local adverse climate and pests (Stanghellini and Montero, 2012; Castilla, 2013). Nevertheless, plastic waste from PC systems pollutes the environment if it is not properly handled, a problem in developing countries where facilities to recycle plastic are not in place.

Today, PC systems have reached such technological levels that greenhouses can be partly or fully automated, relying on increasingly cost-effective and accurate sensors which are connected to a holistic software and corrective equipment steering the climate, irrigation, nutrition and protection of crops within the structure. From light-emitting diodes (LEDs) supplementing sunshine to stimulate plants' growth and development to the use of alternative energy sources (geothermal, solar panels or cogeneration) to operate the greenhouse, the need for highly productive and efficient cropping systems has driven several countries to massively invest in research and development of greenhouse horticulture (Maida, 2017). With labour being one of the largest indirect costs in greenhouse crop production, countries such as the Netherlands have been leading the way in automation, with technologies to transplant, water, grade, sort, and transport plants, as well as control the environment. However, total automation may not be a feasible option for regions such as the SSA, particularly, for smallholders with limited capacity to invest and maintain high technology models. Another method, indoor farming – the practice of producing food in vertically stacked layers that are installed in a warehouse, a skyscraper or a container – has recently become a popular way of producing fresh produce in urban environments where the right combination of rare arable land, cheap energy, reduced transport costs and increasing demand for locally grown foodstuff has made it profitable. Similarly, hydroponics, a method of growing high-value crops by providing plants grown in a soil-less condition with a sterile nutritive solution, is also gaining importance in the PC world: that way, plants are theoretically protected from soil-borne pathogens, pests and nutrient deficiencies.

The role of protected cultivation systems in a challenging world

In a world facing serious environmental challenges, Stanghellini and Montero in their (2012) report discussed the PC of crops as an option that offers promising possibilities to increase the use efficiency of land and farm inputs (water, pesticides, fertilizers), thereby reducing pressure on natural resources. However, there still is a need to improve greenhouse designs and management that are more environmentally friendly and affordable by smallholders in SSA.

Compared to open-field cultivation, greenhouse technologies enable the collection and treatment of water drained through the growing medium. In addition, greenhouse growers tend to make use of high-precision drip-irrigation systems to optimize water use, thus saving water. It has been demonstrated that hydroponic cultivation of lettuce reduces more than ten-fold the water consumption of conventional production (Barbosa et al., 2015). Some countries, such as the Netherlands and Israel, implement strict regulations in terms of water management and encourage their greenhouse growers to collect rain or grey water, to treat it and to (re-)use it for crop irrigation. Such re-use of water resources is highly relevant in areas where water is becoming a scarce resource. This is in line with the bioeconomy approach emphasizing the re-use and recirculation of resources within a given system and thereby increasing the resource use efficiency.

In addition, the confined nature of a greenhouse means that insects find it difficult to attack crops, enabling a lower use of pesticides. The confined environment further increases the suitability of biological control by retaining beneficial insects and predators of parasites inside greenhouses (van Lenteren, 2000). A life-cycle assessment study in Australia showed that 1 kilogram of tomatoes produced in a high tech greenhouse produced six times more carbon dioxide equivalent than 1 kilogram of tomatoes produced in an open field (Page et al., 2012). Heating and management of plastic waste produced by greenhouse cultivation are hotspots in terms of carbon footprints in northern countries, thus adding to the environmental costs (Boulard et al., 2011; Nanna et al., 2018). Whereas, in the tropical areas, the environmental impacts of greenhouse cultivation are expected to be comparatively less due to the lower need for heat. However, management of plastic and other materials used in PC remains a thorny issue, especially in regions where waste treatment and recycling are limited. Alternative materials to plastic are being developed, but they are not commercially viable as yet.

PC as a potential source of income and livelihoods

In the face of a rural exodus and the ageing of the agricultural populations, many developing and developed countries are encouraging youth and women to orient their career towards the agricultural sector (Stone et al., 2005; Alawode, 2016). Through national programmes (financial incentives, increased accessibility to land, technology, training and communication) and/or partnerships with international investment funds, research and extension organizations or non-profit organizations focusing on agricultural development and fighting rural poverty, strategic efforts are being made to keep agriculture an attractive source of jobs and income. As fresh and processed horticultural products are gaining increasing domestic and international demand, it is claimed that fruit and vegetable cultivation will contribute to poverty alleviation and economic development (Weinberger and Lumpkin, 2005; Schreinemachers et al., 2018). Indeed, countries such as Kenya, the Ivory Coast, South Africa and Tanzania are successfully exploiting the niche offered by commercial horticulture, although in different ways, which results in increased foreign income, creation of jobs for skilled and semi-skilled workers, and improved livelihoods for the urban and rural workers (Minot and Ngigi, 2004; Justus and Yu, 2014). This is very much in line with another important principle of bioeconomy: to promote employment and the rural economy and contribute to several of the UN Sustainable Development Goals (SDGs) (Box 7.1).

The Food and Agriculture Organization (FAO), the Technical Centre for Agricultural and Rural Cooperation and International Fund for Agricultural Development (2014) recommended investing in youth living in rural areas, who account for a large percentage of the rural population in developing countries, to enhance agricultural productivity, boost rural economies and ensure food security. As an innovative form of agriculture to produce off-season vegetables when market prices are higher, PC can offer

Box 7.1 The relevance of protected cultivation (PC) in addressing some of the UN SDGs

SDG no. 1: *No Poverty*. By providing food and job opportunities in rural and suburban areas and in territories enduring harsh climatic conditions, PC helps ensure stable revenues to the most vulnerable populations.

SDG no. 2: *Zero Hunger*. Increased yields and production year-round thanks to PC will contribute to food security at family and regional levels.

SDG no. 3: *Good Health and Well-being*. PC offers the possibility to grow nutritious crops such as vegetables throughout the year, regardless of season, thereby fighting malnutrition. Moreover, the reduction of treatment by pesticides, compared to conventional farming, goes towards food safety enhancement.

SDG no. 5: *Gender Equality*. PC empowers women by giving them access to better revenues. Less tedious working conditions through PC can help them achieve a respectable and recognized position in their society.

SDG no. 6: *Clean Water and Sanitation*. The technological development of PC in developed countries has demonstrated its capacity to optimize water consumption through precision irrigation and by collecting, treating and recycling drained water into the irrigation system of the farm. Thus, it ensures the water resource stays clean and available to local populations.

SDG no. 8: *Decent Work and Economic Growth*. PC is subject to rapid technological progress, making horticulture less tedious and more attractive to youth. PC also enables the horticultural sector to be structured, thereby supporting the creation of permanent jobs, the development of agriculture, the reduction of food imports and boosting national economic growth.

SDG no. 11: *Sustainable Cities and Communities*. As megacities emerge, transport of foodstuff from rural to urban areas becomes environmentally and economically unsustainable. PC has found its way into cities via indoor or rooftop farming, creating a new source of jobs there and giving urban dwellers access to locally grown food.

SDG no. 12: *Responsible Consumption and Production*. PC has the potential to become a clean food production system. Technological advances have made it possible to completely control the environment within a greenhouse (therefore limiting the use of pesticides) and to power them with renewable energies.

SDG no. 13: *Climate Action*. By providing job opportunities, food supply, and technological means to produce foodstuff in an environmentally friendly way, PC plays a role in curbing deforestation, reducing the heavy consumption of fossil fuels and food imports, as factors that contribute to climate change.

them meaningful economic opportunities and attractive environments in rural areas.

In this context of the rising importance of horticulture, PC of horticultural products has a key role to play. Pardossi et al. (2004) state that PC may provide good opportunities for jobs in regions of Africa where outmigration is happening at an unprecedented rate. They further emphasize the job creation power of protected horticulture by recalling the success story of the formerly disadvantaged Almeria province in Spain. This hot semi-arid region underwent an astonishing economic growth and the employment of around 80,000 people in this sector in the last two decades, thanks to the development of PC dedicated to the export of vegetables and fruits in the European Union.

A worldwide development of protected cultivation with regional disparities

In terms of numbers, protected horticulture is developing fast on a worldwide scale. The surface of land under protective structures is increasing annually by more than 10 per cent (*HortiDaily*, 2016). However, its worldwide development shows regional disparities in terms of technologies implemented and land used for protected horticulture. Most of the surface increase under greenhouse cultivation has taken place in Asian and Western countries, where the latest greenhouse technologies were developed and applied. The SSA lags behind and, for the time being, there is limited data from this region to assess the current development patterns, constraints and limitations faced by its growers in order to plan for future development.

While greenhouse horticulture in North Africa focused mainly on fruits and vegetables to be exported to Europe, East Africa tends to grow and export mostly flowers (Wageningen University and Research Institutes, 2015). Large commercial farms using PC systems to produce high-value crops for export have sprouted up on the high plateaus of eastern Africa, taking advantage of a climate made congenial by the altitude (sunny and mild) for flower and vegetable production. Nevertheless, their contribution to the technology and knowledge transfer to African farmers needs further study. In most cases, the locals are hired on a daily wage basis, and it is often women who work in these places.

Nowadays, two broad kinds of market-oriented cultivation coexist in SSA regions: (1) greenhouse production dedicated to export markets, and (2) rainfed open field productions catering mainly to the local markets. Although the tomato yield grown in an open field should vary between 60 and 80 tons/ha, it remains well under 20 tons/ha on average in most SSA countries, thus showing a large gap between potential and actual yield. It is worth noting that these tomato yields are far behind the 550 t/ha claimed by greenhouse producers in the Netherlands who do not benefit from the favourable climatic conditions. There is tremendous scope to boost yields in the African context with small improvements in technology, training, and using low cost environmentally friendly materials to construct PC structures.

As the majority of African growers are smallholders producing food for household consumption with a few servicing the domestic market, the techniques and technologies used by the large commercial operations are often not suitable for their own situation in terms of economics, know-how and access to appropriate equipment (Nordey et al., 2017). Thus, the technology must be customized to suit the investment capacity, climatic conditions and markets reached by African growers.

The SSA is characterized by its young and fast-growing population, but paradoxically, the average age of African farmers is 60 years (FAO and ECOSOC, 2014). It has many underused or unexploited fertile lands and an agriculture sector that is still highly affected by climatic and biotic factors, while many Africans still suffer from malnutrition and financial hardship. According to Schreinemachers et al. (2018), PCV is a proven way to improve food security and safety, increase the revenues of farmers and attract youth into the horticultural sector. As previously described, PC can help solve some of the production constraints linked to fruit and vegetable production. To shed more light on the status of PCV in SSA, two case studies from the region are presented.

Case studies from Sub-Saharan Africa

According to FAOSTAT (2019), tomatoes are one of the most widely grown and consumed vegetables throughout the SSA, representing an average proportion of roughly 25 per cent of all vegetables consumed there. Tomato is a key ingredient in many African dishes and a major crop in the SSA where it is sold fresh or processed for longer preservation or for export. Being such an important food item, it is easier to collect data on tomato cropping and marketing than on most other vegetables in the SSA.

Two countries, Tanzania and Benin, underwent large vegetables price variations due to varied rainfall distribution: higher prices are recorded during the main rainy season (around March in Tanzania, and May in southern Benin) because of the lack of control of foliar diseases (late and early blight for tomatoes and sweet peppers, and downy mildew for cucumbers). Similarly, high average temperatures in southern Benin are detrimental to the flowering, pollination and fruit set of tomato plants. Celebrations such as Ramadan or Christmas and New Year's Eve also influence price increases due to a higher demand for various foods. Thus, tomato prices on the market vary greatly and can increase up to sevenfold, for e.g. US$0.5–3.4 per kg in Benin (Cotonou) as compared to US$0.2–0.6 per kg in Tanzania (Arusha).

In such varied supply demand market conditions, it is of economic interest for producers to sell tomatoes at the times of the year when prices are high, during the off-season or at celebrations. To do so, growers need to adapt their cropping systems to harvest at the opportune times. Protected cultivation of tomato can enable farmers to grow tomatoes even when environmental conditions are adverse and not suitable for cultivation outdoors. Table 7.1 shows the local contexts which farmers willing to implement PC of tomatoes need to develop.

The high price of imported greenhouses coupled with soil-borne diseases such as bacterial wilt or nematodes and the adverse effects of high temperatures on tomato plants complicate farmers' evolution to off-season PC of tomatoes. The success rate of grafting tomato plants decreases quickly above 21 °C, whereas flowering, pollination and the fruit set of tomatoes are limited above 28 °C, which are the two major constraints faced by Beninese farmers.

Table 7.2 clearly points out the increase in both average yield and average production cost per crop changing from open field cropping to greenhouse farming. But most importantly, the estimated average gross margin per crop is 4 times higher in Benin and nearly 10 times higher in Tanzania for greenhouse farming than for open field production, which clearly demonstrates the profitability of PC of tomatoes, provided there is a market for high quality produces (Figure 7.2). Nevertheless, it is worth

Table 7.1 Agro-climatic context for greenhouse cultivation

Column 1	*Benin (coastal region)*	*Tanzania (Inland plateau)*
Climatic conditions	Tropical wet and dry climate Two distinct rainy and dry seasons per year. Average annual temperature: 28°C Average annual relative humidity: 83%. Low yearly temperature and relative humidity amplitudes	Subtropical highland climate Two distinct rainy and dry seasons per year Average annual temperature: 18°C Average annual relative humidity: 70%. High day-night temperature amplitudes
Main production impediments	Heavy rains, diseases and pests, high temperatures	Heavy rains, diseases and pests
Average cost of a greenhouse* ($/m²)	10–35	10–30
Commercial greenhouse	28–30	25
Local craftsman	8–13	10

Source: Based on field data.

Notes

*Unequipped, metallic structure roof-covered with polyethylene and walls covered with insectproof net.

Table 7.2 Technical and economic data on tomato cropping

Tomato crop	*Benin (coastal region)**		*Tanzania (inland plateau)*	
	Open field	*Greenhouse*	*Open field*	*Greenhouse*
Yearly national production per crop (t/year)	339,560	NA	565,441	NA
Average yield per crop (t/ha)	8.41	57.5	18.5[a]	100–120[a]
Average production costs per crop ($/m²)	0.25	1.07	0.09[a]	0.3[a]
Average gross margin per crop ($/m²)	0.148	0.81	0.17[a, b]	1.6[a, b]

Source: adapted from Ahouangninou (2013); Mensah (2016).

Notes
[a] Personal estimations; [b] Cost-benefit analysis greenhouse cultivation Kenya.

Figure 7.2 Tomatoes grown under a high tunnel in Arusha, Tanzania.
Source: photo by T. Nordey.

noting that the temperate climate on the Tanzanian and Kenyan highlands is suitable for PC, allowing year-round production of vegetables, although warmer temperatures in Benin restrict vegetable production under simple shelters to the cool season (Gogo et al., 2014; Simon, 2014; Mensah, 2016).

Factors affecting the development of PCV in Sub-Saharan Africa

Several projects and much research have been conducted in many SSA countries, including Kenya, Tanzania, Rwanda, Senegal and Benin, to promote PCV. Results stress the interest of using nets to increase the yield of several vegetable crops, including tomatoes, French beans, or cabbages while reducing the use of pesticides (Martin et al., 2006; Licciardi et al., 2007; Gogo et al., 2012; Nordey et al., 2017). This section of the chapter intends to cover the main constraints and limitations these projects have faced.

The land factor and the investment scale

Besides climate and market, price and availability of land are factors influencing the agricultural development in the region. In parts of the European Union, for instance, expensive and limited land has driven farmers towards agricultural intensification to improve yields via an increase in resource use efficiency and better control of the environmental factors. Consequently, this has encouraged the development of PC for high-value crops. In contrast, cheap and abundant land tends to orient farmers towards extensive cropping systems: in fact, the production increase in SSA is assumed to be primarily due to the rise of the total cultivated area rather than to the augmentation of yields (Dubois, 2011). These circumstances do not necessarily prompt farmers or agribusinesses to set up protected cropping systems, but rather to expand their land when possible. Moreover, the major greenhouse and PC equipment is produced by Western and Asian companies. In SSA, the equipment used in PC systems is non-existent, inadequate or more expensive as it is generally imported, preventing smallholder farmers from practising protected horticulture over large areas.

Likewise, commercial banks in many SSA countries are still reluctant to invest in the agricultural sector, since they consider it risky (due to high weather variability and influence, the inability of farmers to provide the necessary collateral and the risk involved in repayment of loans) and administratively costly (Madugu and Bzugu, 2012; Baba et al., 2015). Also, agricultural insurance meeting the risk management needs of farmers is still too underdeveloped to convince the banks otherwise (Hess et al., 2015). Consequently, the share of agricultural credits in total bank lending is relatively low in SSA and the banks' loan offers often come with double-digit interest rates. Therefore, the capital required to establish a farm with protected cropping systems in SSA is relatively substantial for an African smallholder farmer.

Figure 7.3 Cabbage under net in southern Benin.
Source: photo by A. Mensah, 2019.

Yet, comparing small netted structures with open field cabbage crops in Benin, Vidogbéna et al. (2015) found a higher cost-benefit ratio for netted protection than unnetted protection. They also reported that nets provided more stable costs and yields. More stable cash flows and a decrease in natural risks caused by biotic and abiotic factors, as the benefits of PC systems, could encourage banks to invest more in protected horticulture.

Although the effectiveness of physical protection is tested in southern Benin, the high cost of insect nets and especially their lack of availability are factors limiting the use on large areas. For many crops to be transplanted, physical protection is more noted at the nursery level with the use of nets of different mesh size, depending on the levels of ventilation and pests' protection desired. The same study by Vidogbéna et al. (ibid.) also showed that it was the premium sale price allowed by the higher quality of the harvested cabbages under protective structures that made protected systems more profitable (Figure 7.3).

Market access and policies

Incentives to adopt PC systems differ among states. European farmers, for instance, are subject to strict regulations regarding the use of pesticides and therefore growers tend to resort to the physical protection of their crops and to integrated pest management methods as alternatives. To help farmers make this transition, some European states, such as France or the Netherlands, offer loans, grants or subsidies to build, expand, renovate or equip greenhouses (Breukers et al., 2018; Xicluna, 2019). This is in accordance with the general growing concern among European customers regarding food origin and safety, especially about pesticide residues in foodstuffs, and the impacts of pesticides and food transport on their natural environment. The result is a demand for more precise traceability, closer scrutiny of farmers' practices and the development of new markets (organically and/or locally grown products). Those fast-emerging markets are opportunities for farmers to implement protected cultivation as they entail higher consumer prices that enable them to amortize their protective equipment and production costs. Various actors along the product value-chains are well connected and hence it is possible for farmers to produce for the target markets, rather than produce and hope that the market will absorb their production at reasonable prices. Such organized value chains with assured markets help producers to plan their investments.

On the contrary, value chain actors are not strongly linked and there exist wide gaps and weaknesses in the chains, making it difficult for smallholders in SSA to operate. Not only are their resources, investments, and technology limited, but they are also constrained by unstable prices and inefficient markets. Indeed, the lack of segmentation into conventional and organic fruits and vegetables in these markets means that products from PC are not yet profitable, although a niche of consumers of these products is emerging.

Regulations on pesticides used in the SSA tend to be more permissive or face difficulties in being enforced. Although consumers focus primarily on food appearance, availability and affordability, awareness of the health hazards caused by pesticide residues in food products is increasing. Middle- and high-income consumers tend to be more willing to pay more for safe vegetables, as shown in a study conducted in Benin and Ghana by Nouhoheflin et al. (2005). Nevertheless, in a context where wholesalers and consumers generally make little distinction in quality between traditionally grown and organic or protected cropping systems' products, farmers are not encouraged to look for more expensive alternatives than pesticides to protect their crops.

Agronomic issues induced by greenhouse production

Shelters such as greenhouses or tunnels are mainly built in the tropics to protect crops from insect pests and rainfall. Yet, in practice, shelters are not perfectly sealed and although they slow down the penetration of insect pests, they might paradoxically trap predators of the crops as well. Indeed, pests may take advantage of suitable climatic conditions and the absence of natural enemies to spread faster than in the open field.

In the tropics, higher temperatures under covers can be detrimental to crop development. In these regions, plastic nets are used on the shelter's walls or all over to improve air ventilation and thereby reduce the increase in temperature. Nevertheless, the optimum type of nets to use must be found, since the larger the mesh and the vents, the better the ventilation and air renewal within the crop, thus reducing risk of fungal diseases, but the higher the risks of intrusion by small insect pests. Farmers need to be trained on the different materials and how to build tunnels that are insect-proof.

As protective structures constitute permeable physical barriers, they tend to retain the water vapour coming from the crop's transpiration and soil evaporation. This results in a higher relative humidity within the shelter that is suitable for fungal disease development (Blancard and Laterrot, 2009). Relative humidity can be controlled through passive ventilation (naturally, using nets and vents) or active ventilation (using fans), the latter being more efficient but entailing higher cost of equipment of energy consumption. Photo-selective covers (to block the ultra-violet (UV) rays) hampering the growth of fungal diseases are being developed (Thind and Schilder, 2018) but their absence on the African market, and their prices make them out of reach of the common smallholder farmers. Finally, soil-borne diseases are a major problem in PC since crop rotation is reduced to high value crops, i.e. tomato, cucumber, pepper, eggplant, for the sake of profitability.

Ireri et al. (2018) conducted a study on Kenyan farmers' knowledge and experience of practices to control soil-borne pathogens (bacterial wilt and root-knot nematode) as part of the protected cultivation of tomato. The investigation showed that: (1) most of the tunnels were used for less than two years of production; (2) about 62.5 per cent of the farmers taking part to the study faced issues with soil-borne disease; and (3) crop losses amounted to 50–100 per cent.

With the phasing out of soil fumigation, few alternatives remain to overcome soil-borne diseases under shelters, excepted soil-less cultivation and grafting. However, those two techniques entail a sharp increase in production costs for the smallholder farmers.

Rethinking the path towards protected cultivation

Protected cultivation has found its way into the agricultural sector of many countries, especially Western and Asian regions. Nevertheless, it implies financial, commercial and agronomical conditions and constraints that can be unattainable and prohibitive to African smallholder farmers.

It is obvious that there is a gap between the actual levels of technology of cropping systems of smallholder farmers and greenhouse technologies. Previous experiences in upscaling protected cultivation techniques stressed the need to provide technical support to help farmers adapt their whole practices. Proper management of irrigation, fertilization, varieties, and crop pests are essential prerequisites to benefit from greenhouse cultivation. These simple and cost-effective techniques and technologies figure as first steps that could help African growers improve their crops' yield potential. In this regard, the room for improvement is enormous.

A drip-irrigation system in East Africa costs less than $1.5/m^2 but paradoxically, only 4 per cent of agricultural land is irrigated in SSA versus 37 per cent in Asia and 18 per cent worldwide (Burney et al., 2013). Seeds of improved varieties including hybrids are now available throughout the African continent, developed and distributed by national and international research centres as well as local distributors or international seed companies, but old open-pollinated varieties are still widely used in SSA (Fufa et al., 2011). Huge improvements are also expected in the control of insect pests in SSA and in the sanitary quality of production through the better use of pesticides (de Bon et al., 2014).

Likewise, the large post-harvest losses reported in SSA which can be as high as 43.5 per cent for leafy vegetables and 33.7 per cent for tomatoes (Affognon et al., 2015) raise some questions about the relevance of increasing production without improving access to markets and equipment for storage and transport. Such observations confirm the need to consider with care how PC techniques can fit with actual cropping systems and with the constraints of the local value chains. Efforts are also required to adapt techniques to local climatic conditions and investment capacities. Innovative and entrepreneurial farmers do not wait for solutions from research and are building improvised structures using organic residues (straw, wood chips) as mulch and small farm-made shelters with recycled mosquito nets to produce seedlings or to protect their crops. The establishment of a business ecosystem between young local entrepreneurs and producers would be an asset for the large-scale dissemination of physical protection. It is interesting to note that successes with simple shelters encourage farmers to pursue this mode of production with the construction of higher forms of shelter.

Conclusion

The protected cultivation of vegetables has the potential to be a significant part of the global bioeconomy development. It can directly contribute to achieving several Sustainable Development Goals, making it a tool to be integrated into national and international development strategies. In that respect, PC is already encouraged and recognized as an approach to promote sustainable agriculture development, in many regions of the world. However, in SSA, it needs more favourable policies to scale it up for the benefit of smallholders. State subsidies of greenhouses or shelters in developing countries could contribute to the adoption of this technology. However, promoting the development of PC to smallholder farmers without considering basic agricultural techniques such as irrigation, fertilization, improved varieties, pest management and stable markets would be like putting the cart before the horse. Most of this improvement can only happen if the knowledge, access to equipment, market opportunities, and financial resources are brought to the farmers' level. It is only then that farmers will take advantage of the potential of protected cultivation techniques.

References

Affognon, H., Mutungi, C., Sanginga, P. and Borgemeister, C. (2015) 'Unpacking postharvest losses in Sub-Saharan Africa: A meta-analysis', *World Development*, vol. 66, pp. 49–68.

Ahouangninou, C.C.A. (2013) 'Durabilité de la production maraichère au sud-Bénin: un essai de l'approche écosystémique', University of Abomey-Calavi, Abomey-Calavi, Bénin.

Alawode, O. (2016) 'Attracting young people to farming'. Available at: https://spore.cta.int/en/production/all/article/attracting-young-people-to-farming-sid-03d5afc22-fecb-46ee-8b68-8129ffc6eb5e

Baba, H., Abdallah, A.-H. and Hudu, Z. (2015) 'Factors influencing agricultural credit demand in Northern Ghana', *African Journal of Agricultural Research*, vol. 10, no. 7, pp. 645–652.

Barbosa, G., Gadelha, F., Kublik, N., Proctor, A., Reichelm, L. et al. (2015) 'Comparison of land, water, and energy requirements of lettuce grown using hydroponic vs. conventional agricultural methods', *International Journal of Environmental Research and Public Health*, vol. 12, no. 6, pp. 6879–6891.

Blancard, D. and Laterrot, H. (2009) *Les maladies de la tomate: Identifier, connaître, maîtriser*. Versailles: Quae.

Boulard, T., Raeppel, C., Brun, R., Lecompte, F., Hayer, F., et al. (2011) 'Environmental impact of greenhouse tomato production in France', *Agronomy for Sustainable Development*, vol. 31, no. 4, pp. 757–777.

Breukers, A., Hietbrink, O. and Ruijs, M. (2018) 'The power of Dutch greenhouse vegetable horticulture: An analysis of the private sector and its institutional framework'. Available at: http://edepot.wur.nl/27928

Burney, J.A., Naylor, R.L. and Postel, S.L. (2013). 'The case for distributed irrigation as a development priority in sub-Saharan Africa', *Proceedings of the National Academy of Sciences of the U.S.A.*, vol. 110, no. 31, pp. 12513–12517.

Castilla, N. (2013) *Greenhouse Technology and Management*. London: CABI.

de Bon, H., Huat, J., Parrot, L., Sinzogan, A., Martin, T., et al. (2014) 'Pesticide risks from fruit and vegetable pest management by small farmers in sub-Saharan Africa: A review', *Agronomy for Sustainable Development*, vol. 34, no. 4, pp. 723–736.

Dubois, O. (ed.). (2011) *The State of the World's Land and Water Resources for Food and Agriculture: Managing Systems at Risk*. New York: Earthscan.

FAO (Food and Agriculture Organization) (2014) *Youth and Agriculture: Key Challenges and Concrete Solutions*. Rome: FAO.

FAO and ECOSOC (2014) 'Sustainable urbanization'. Available at: www.un.org/en/ecosoc/integration/pdf/foodandagricultureorganization.pdf

Food and Agriculture Organization of the United Nations, Technical Centre for Agricultural and Rural Cooperation, and International Fund for Agricultural

Development (2014) 'Youth and agriculture: Key challenges and concrete solutions', Rome: FAO.

FAOSTAT (2019) 'Tomato consumption in Africa'. Available at: www.fao.org/faostat/en/#data/FBS

Fufa, F., Hanson, P., Dagnoko, S. and Dhaliwal, M. (2011) 'AVRDC – The World Vegetable Center tomato breeding in Sub-Saharan Africa: Lessons from the past, present work, and future prospects', *Acta Horticulturae*, no. 911, pp. 87–98.

Gogo, E.O., Opiyo, A., Ulrichs, C. and Huyskens-Keil, S. (2014) 'Postharvest treatments of African leafy vegetables for food security in Kenya: A review', *African Journal of Horticulture Science*, vol. 9, pp. 32–40.

Gogo, E.O., Saidi, M., Itulya, F.M., Martin, T., and Ngouajio, M. (2012) 'Microclimate modification using eco-friendly nets for high-quality tomato transplant production by small-scale farmers in East Africa', *HortTechnology*, vol. 22, no. 3, pp. 292–298.

Graamans, L., Baeza, E., van den Dobbelsteen, A., Tsafaras, I., and Stanghellini, C. (2018) 'Plant factories versus greenhouses: Comparison of resource use efficiency', *Agricultural Systems*, vol. 160, pp. 31–43.

Hess, U., Hazell, P. and Kuhn, S. (2015) 'Innovations and emerging trends in agricultural insurance', paper presented at the G20 Round-Table on Innovations in Agricultural Finance, September, Antalya, Turkey.

HortiDaily (2016) 'Global greenhouse vegetable area increased 14% over 2015'. Available at: www.hortidaily.com/article/6023258/global-greenhouse-vegetable-area-increased-14-over-2015/

Ireri, D.F., Murungi, L.K., Ngeno, D.C. and Mbaka, J. (2018) 'Farmer knowledge of bacterial wilt and root-knot nematodes and practices to control the pathogens in high tunnel tomato production in the tropics', *International Journal of Vegetable Science*, vol. 25, no. 3, pp. 1–13. https://doi.org/10.1080/19315260.2018.1499690.

Justus, F. and Yu, D. (2014) 'Spatial distribution of greenhouse commercial horticulture in Kenya and the role of demographic, infrastructure and topo-edaphic factors', *ISPRS International Journal of Geo-Information*, vol. 3, no. 1, pp. 274–296.

Lamont, W.J. (2005) 'Plastics: Modifying the microclimate for the production of vegetable crops', *HortTechnology*, vol. 15, no. 3, 477–481.

Licciardi, S., Assogba-Komlan, F., Sidick, I., Chandre, F., Hougard, J. M. and Martin, T. (2007) 'A temporary tunnel screen as an eco-friendly method for small-scale farmers to protect cabbage crops in Benin', *International Journal of Tropical Insect Science*, vol. 27, nos. 3–4, pp. 152–158.

Madugu, A.J. and Bzugu, P.M. (2012) 'The role of microfinance banks in financing agriculture in Yola North Local Government Area, Adamawa State, Nigeria'. Available at: www.academia.edu/7523311/The_Role_of_Microfinance_Banks_in_Financing_Agriculture_in_Yola_North_Local_Government_Area_Adamawa_State_Nigeria

Maida, J. (2017) 'Global greenhouse horticulture market 2017–2021: Key geographies and forecasts by Technavio Research'. Available at: www.businesswire.com/news/home/20170713006114/en/Global-Greenhouse-Horticulture-Market-2017-2021-Key-Geographies

Malézieux, E., Crozat, Y., Dupraz, C., Laurans, M., Makowski, D., Ozier-Lafontaine, H. and Valantin-Morison, M. (2009) 'Mixing plant species in cropping systems: Concepts, tools and models. A review', *Agronomy for Sustainable Development*, vol. 29, no. 1, pp. 43–62.

Martin, T., Assogba-Komlan, F., Houndete, T., Hougard, J.M. and Chandre, F. (2006) 'Efficacy of mosquito netting for sustainable small holders' cabbage production in Africa', *Journal of Economic Entomology*, vol. 99, no. 2, pp. 450–454.

Mensah, A. (2016) 'Intensification de la culture de tomate sous abri couvert de filet anti-insectes en région chaude et humide du Sud-Bénin', *Science et Technique – Revue Burkinabé de La Recherche, Sciences Naturelles et Agronomie*. Symposium International sur la Science et la Technologie (SIST 2016) (Spécial hors série n° 2).

Minot, N. and Ngigi, M. (2004) 'Are horticultural exports a replicable success story?: Evidence from Kenya and Côte d'Ivoire'., Washington, DC: International Food Policy Research Institute.

Nanna, M., Batista, M.T., Baptista, F.J., Schettini, E. and Vox, G. (2018) 'Mapping greenhouse plastic wastes in the west region of Portugal', *Acta Horticulturae*, vol. 1227, pp. 257–264.

Nordey, T., Basset-Mens, C., De Bon, H., Martin, T., Déletré, E., Simon, S. and Malézieux, E. (2017) 'Protected cultivation of vegetable crops in sub-Saharan Africa: Limits and prospects for smallholders: A review', *Agronomy for Sustainable Development*, vol. 37, no. 6. Available at: http://publications.cirad.fr/une_notice.php?dk=585832

Nouhoheflin, T., Coulibaly, O., Cherry, A.J., Al-Hassan, R.M. and Adegbola, P. Y. (2005) 'Consumers' perceptions and willingness to pay for organic vegetables in Benin and Ghana', *African Association of Agricultural Economists-Conference Proceedings*, no. 1, pp. 1–11.

Page, G., Ridoutt, B. and Bellotti, B. (2012) 'Carbon and water footprint tradeoffs in fresh tomato production', *Journal of Cleaner Production*, vol. 32, pp. 219–226.

Pardossi, A., Tognoni, F. and Incrocci, L. (2004) 'Mediterranean greenhouse technology', *Chronica Horticulturae*, vol. 44, no. 2, pp. 28–34.

Schreinemachers, P., Simmons, E.B. and Wopereis, M.C.S. (2018) 'Tapping the economic and nutritional power of vegetables', *Global Food Security*, vol. 16, pp. 36–45.

Simon, P.W. (2014) 'Progress toward increasing intake of dietary nutrients from vegetables and fruits: The case for a greater role for the horticultural sciences', *Hortscience*, vol. 49, no. 2, pp. 112–115.

Stanghellini, C. and Montero, J.I. (2012) 'Resource use efficiency in protected cultivation: Towards the greenhouse with zero emissions', *Acta Horticulturae*, vol. 927, pp. 91–100.

Stone, G., Coutts, J., Casey, M. and Coutts, A. (2005) 'Strategy to attract young people to horticulture', *Acta Horticulturae*, vol. 67, pp. 339–345.

Thind, T.S. and Schilder, A. C. (2018)''Understanding photoreception in fungi and its role in fungal development with focus on phytopathogenic fungi', *Indian Phytopathology*, vol. 71, no. 2, pp. 169–182.

van Lenteren, J.C. (2000) 'A greenhouse without pesticides: Fact or fantasy?', *Crop Protection*, vol. 19, no. 6, pp. 375–384.

Vidogbéna, F., Adégbidi, A., Assogba-Komlan, F., Martin, T., Ngouajio, M., Simon, S. and Parrot, L. (2015) 'Cost:benefit analysis of insect net use in cabbage in real farming conditions among smallholder farmers in Benin', *Crop Protection*, vol. 78, pp. 164–171.

Wageningen University and Research Institutes (2015) 'Protected horticulture in Africa'. Available at: www.wur.nl/en/Research-Results/Research-Institutes/plant-research/greenhouse-horticulture/Research-themes/International/Subthemes-International/Africa.htm

Weinberger, K.M. and Lumpkin, T.A. (2005) 'Horticulture for poverty alleviation: The unfunded revolution'. Available at: https://papers.ssrn.com/abstract=781784

Xicluna, P. (2019) 'Plan de compétitivité et d'adaptation des exploitations agricoles@ Alim'agri'. Available at: https://agriculture.gouv.fr/plan-de-competitivite-et-dadaptation-des-exploitations-agricoles

8 Mediterranean anthroscapes

A bioeconomy domain for sustainable land use

Erhan Akça, Suha Berberoğlu, Takanori Nagano and Selim Kapur

Introduction

Turkey sits at the crossroads of two continents, with most of its Asian territory originating in ancient Anatolia, and Thrace originating in the European part. The vast geological formations and climatic regions yield a very rich flora, close to 12,000 species. The country, which is approximately 800,000 km^2, is extensive but the average altitude of 1,100 m, steep topography and uneven precipitation throughout the country place certain limitations on crop production. Moreover, the semi-arid forests, excluding the Black Sea region, with more than 1,000 mm annual precipitation and highly degraded grazelands are another drawback for dense vegetation cover. The unfavourable geographical conditions in the central, eastern and south-eastern parts of the country have led to the concentration of the population mostly in the coastal areas, which puts high pressure on natural resources. Anatolia was never a territory rich in fossil fuels and for millennia its settlers relied on wood and shrubs to satisfy their fuel demand, which historically caused severe erosion, particularly in areas with relatively high biomass production, such as the Mediterranean coast (Akça et al., 2009).

Turkey imports more than 75 per cent of its energy demand, which is met by fossil fuels. And its economy, in every sector from transportation to heating and from energy to agriculture, is dependent on a fossil-fuel-based development pathway (Şahin, 2018). The country's energy import costs are also increasing amid fluctuating oil prices that amounted to more than \$37 billion or nearly 16 per cent of its total import value in 2017 (Şengül, 2017). Thus, the country needs alternative renewable sources to meet its increasing energy demands. Turkey established its first major national renewable energy targets in its 2010–2014 Strategic Plan. The Plan included the goal of replacing fossil fuels with renewables, based on solar and wind sources by 30 per cent of total energy production by 2023. However, bio-fuel was not mentioned in the Strategic Plans because agricultural lands would not be very productive unless irrigated and had mainly been used for food and fibre production since the foundation of the republic in 1923. It is necessary to take a closer look at the Anatolia region, its land resources and actual potential to support the new renewable energy targets, especially by boosting the use of bio-fuel or developing other sectors to generate bio-resources. Thus, this chapter analyses the potential of bioeconomy in the Anatolia region based on sustainable use of resources.

Anthroscapes of Anatolia

The Mediterranean Basin, which has cradled many cultures, such as the Phoenicians, the Greeks, the Romans, the Arabs and the Ottomans since Antiquity, has been the area for several land use activities where people developed innovative land management practices, including water harvesting and terracing to conserve limited water and soil and to gain maximum biomass output to meet the increasing demand for resources. We call this historical human-reshaped landscape the 'anthroscape'. The anthroscapes differ from natural landscapes not only topographically but also in their energy and material flows. The anthroscape context includes an amalgamation of present and ancient land use practices and knowledge which blossomed from the combination of topography, natural resources, biodiversity and climate of the region. Quantification of the entire anthroscape is not a simple task, and in early times it generally depended on observations and assumptions rather than measurements (Jacobsen and Adams, 1958). Contemporary anthroscape quantifications have generally focused on the magnitude of degradation, not their productivity or sustainability.

Humans did not need to modify the environment in earlier times as the resources readily met their basic needs. However, to serve the needs of the growing populations, man started to shape the environment because Palaeolithic hunting, gathering and fishing did not provide enough food. One of the first modifications of the environment was most probably slash and burn cultivation that emerged 11,000 years ago in the arid/semi-arid Middle East, which is now the territories of Jordan, Syria, Israel and Turkey (Lev-Yadun et al., 2000). In these early settlements, the first farmers cultivated emmer, einkorn wheat, barley and legumes in the eighth millennium bc, along with grazing to support animal husbandry (Rasmussen, 2010). Farming of crops and livestock thus forced them to settle down, a big change from being hunters and gatherers. To irrigate crops in the arid Middle East, Sumerians employed the waters of the Euphrates and Tigris Rivers in the sixth millennium bc (Murphy, 2007). This permitted the relatively large populated Mesopotamian cities to flourish. Ever since, humans have been able to structure their environment according to their needs anywhere in the world. The land transitions around the world created long-standing sustainable domains such as paddy fields in East Asia, olive orchards and vineyards in the Mediterranean Basin and grazing lands in Anatolia (Akça et al., 2008), which were termed anthroscapes by

Eswaran et al. (2005). The anthroscape scale varies from region to region, it is a region where commodities are produced for human self-sufficiency. In the Mediterranean context, it stretches from the coast to the high mountain ranges (Figure 8.1).

Centuries were classified in many ways by human achievements such as the Bronze Age, the Iron Age, and the twentieth century, the age of innovation. However, the twenty-first century will be remembered as the age of extinction due to destructive global land degradation and desertification, climate change and biodiversity loss. Humans shaped the environment in a sustainable way until the eighteenth century apart from a few areas such as the saline areas in Mesopotamia and the Indus Valley, and in the Mayan lands where the soil was depleted of nutrients. Thus, Eswaran et al. (2010) suggested that anthroscapes might also embrace human-induced land degradation, with major negative changes leading to a decline in land quality. Lowdermilk (1950) reported an excellent survey investigating how extinct cultures were based in degraded lands around the world and pinpointed a 'hundred dead cities' (Mattern, 1944) which once were wealthy centres.

In recent times, the success of environmental management is measured by its organic carbon dynamics, particularly, by the net primary production (NPP) (Erşahin et al., 2017). The United Nations Conference on Sustainable Development (Rio+20), in June 2012, recognized organic carbon along with land productivity and land use as one of the indicators of land degradation neutrality (Chasek et al., 2015), which is a common approach used to evaluate land functionality. Although net primary productivity (NPP) describes the functions of soil, water and vegetation components of the ecosystem (Malhi et al., 2011), the NPP measurement alone cannot reflect the soil quality for productivity, since, in some cases, fresh carbon input from the NPP accelerates the decomposition of the soil organic

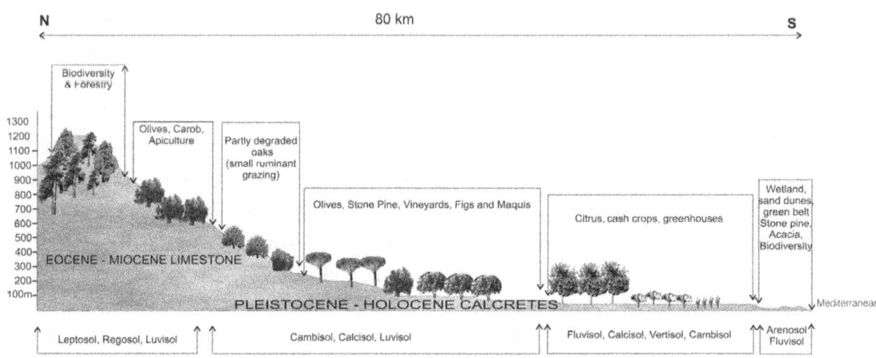

Figure 8.1 The Mediterranean anthroscape of southern Turkey.

Source: adapted from Kapur et al. (2019).

matter through biological activities (Guenet et al., 2018). Hence, both NPP and the soil organic carbon of an ecosystem should be evaluated together to define the biomass production efficiency of the system. Measurement of these two indicators of the ecosystem will reveal the potential of the land's bioeconomy potential. High NPP may be achieved in a short time of two years in an environment with a high level of irrigation and chemical fertilization. However, if fresh input into soil organic carbon pool is not sustained, use of external inputs to supplement carbon may lead to an increase in costs, which could exceed the income from the biomass production. When there is a balance between both NPP and the soil organic carbon pool, land productivity will be sustained for longer periods, which in turn will provide sufficient food and fibre to the communities. Securing food and fibre demands will fulfil the UN's SDGs Goal 1 (eradicate poverty) and Goal 2 (zero hunger), but Goal 15 (protecting life on land) depends on sustainable biomass production.

The Mediterranean anthroscapes have mostly developed on calcretes in the semi-arid part of Turkey and have focused on securing food and fibre production using terraced farming and by saving water since the time of Hittites (2500 bc), the Romans and the Ottomans. The variety of anthroscapes in Anatolia and the primary biomass and livestock production domains have continued to support the food and fibre demand of humans, amid threats of global climate change, land degradation, desertification, loss of biodiversity, and boosted land biomass production. This can be attributed to their high carbon sequestration and water-saving capabilities. Moreover, the anthroscapes, which are natural highlands, have been used for grazing and valleys were used for intense agricultural production. This proves that humans have experience in how to adapt to elevation, slope and soil types in the anthroscapes they have shaped. The present demand for land is far greater though, due to ever increasing agriculture and tourism activities. In 2015, the number of tourists reached over 320 million as opposed to 500 million residents in the Mediterranean Basin. A boost in the tourism sector creates a higher demand for local agricultural products. But it also means more pressure on the local natural resources. Since 1970, the traditional agricultural practices in anthroscapes have been converted to irrigated agriculture, which has boosted local farmers' income. Today, farmers pursuing traditional agriculture in anthroscapes are facing a dilemma of having to choose between preserving the ecological services of the anthroscapes or boosting their income by converting these lands into irrigated agricultural fields. To ensure sustainability in the long run, we need to maintain a balance in agricultural production versus protection of ecological services.

Thus, the study sought to outline carbon zones in the Göksu anthroscape, which mainly extends on calcrete in the Mediterranean region of Turkey, to quantify its organic carbon status which reflects the region's NPP. This can provide an index and guidelines if the region is regarded as a 'bioeconomy domain' (BiD), where carbon sequestration and the ecological footprint should be balanced with income generation. Then a BiD becomes a special zone in semi-arid areas where soil and water management can support relatively high biomass production. In a BiD, soil and water resources do not necessarily have to be of high quality or exist in large amounts, but the crop pattern needs to be well chosen to obtain the maximum possible biomass while maintaining the soil quality and the water resources.

The Göksu Project was funded by the Turkish Scientific and Technical Council to determine the carbon dynamics of a semi-arid anthroscape with a number of limitations, such as shallow soils, steep topography, uneven water resource distribution and population pressure on prime lands that are very scarce in the basin. Although limitations existed, the locals have developed traditional management techniques to secure food and fibre production in the basin. One thing that is lacking in Göksu is the local people's awareness of the bioeconomy approach. The important thing in the Göksu anthroscape (hereafter GA) is to precisely determine the carrying capacity for sustainable land use, involving all stakeholders, which could pave the way for bieconomy planning and development.

This study is one of the preliminary reviews that will help not only to evaluate the bioeconomy potential of the region but also to contribute to increase bioeconomy awareness among policy-makers and land users in the specific environments who want to achieve high biomass production with sustainable environmental management. Turkey considers the agricultural sector only as a source of food and fibre production despite the country's energy dependence on imported fuels. Hence, this review will provide a new perspective to increase agricultural production efficiency and at the same time contribute to renewable energy through sustainable soil management and enhanced soil organic carbon input through appropriate land uses.

Materials and methods

Study area

The Göksu Basin is located in the south-eastern part of the Mediterranean Basin, in Turkey, stretching over 10,000 km^2 of coastal areas to an altitude of 2,100 m above sea level, showcasing a wide range of land uses from World Heritage Areas in the coastal zone to grassland in the highlands. The GA extends from the southern Taurids (2,100 m above sea level) to the Mediterranean Sea of Turkey in 36° 14'20"–37° 13'21" N, and 32° 06'29"–34° 04'39"E (Figure 8.2). The Mediterranean climate dominates the area with an average annual precipitation of 636 mm and a temperature of 19.1 °C. Cambrian, Ordovician, Devonian and Permo-Carboniferous limestones along with Quaternary sediments, are the main geological formations of the GA (Başaran, 2003).

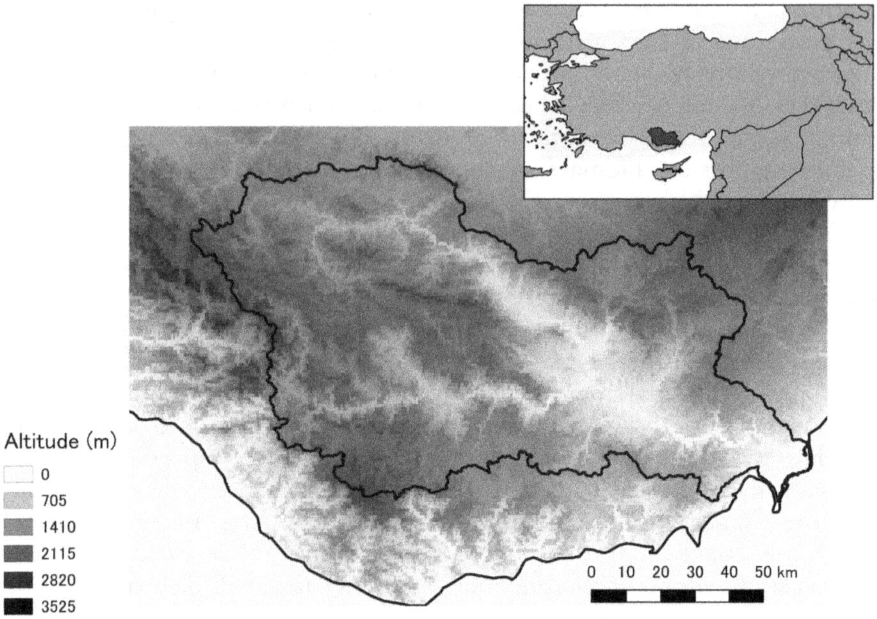

Figure 8.2 Location of the Göksu anthroscape.

The GA has a sloping topography with an average slope of 6.11 °. However, in this it varies from 0 to 89 ° and only 1 per cent of the region is flatland. GA's main soil type is Leptosols, followed in descending order by Cambisol, Calcisol, Kastanozem, Luvisols, Fluvisols and Gleysol and colluvial Cambisol-Calcisols (IUSS WRB Working Group, 2015). For clarification of soil types in GA we used the definitions of the IUSS WRB Working Group (2015). Leptosols are described as very shallow soils on hard rock or highly calcareous materials. They may also be deeper soils but are extremely gravelly and/or stony. They are common soils in the mountainous parts of the GA with natural forests along with pastures. Cambisols are described as the soils at the beginning of soil formation, developed as medium and fine-textured materials derived from a wide range of rocks, mostly in alluvial, colluvial and aeolian deposits. Most of these soils are intensively used for cultivation in the GA. Cambisols are accepted as the most productive soils on Earth in temperate climates. Calcisols are the soils rich in calcium carbonate with a substantial secondary accumulation of lime, and generally with moderate soil depths (30–60 cm), common in calcareous parent materials and in arid and semi-arid environments. Here, the natural vegetation is sparse and dominated by xerophytic shrubs and trees and/or ephemeral grasses. Calcisols in GA are used mostly for olive orchards and pine along with oak forests developed on these soils. Luvisols, with an average soil depth of 30–60 cm in the GA, are soils whose dominant characteristic is a marked textural differentiation within the soil profile, with the surface horizon depleted of clay and with accumulation of clay in a subsurface. These soils are mostly used for olive orchards, vineyards and vegetables in the GA. Kastanozems in the GA are mostly found in grazing lands. They are not common soils but are rich in organic matter, and originally underpinned early maturing native grasslands vegetation, which produces a characteristic brown surface layer in the first metre underground. Fluvisols in GA are found on alluvial plains, river fans, valleys and tidal marshes and are considered young soils. They are deep due to sediments deposited by the water. Fluvisols of the GA are the most productive soils for vegetable production. Gleysols are only found in marshes in the GA and are soils that, unless drained, are saturated with groundwater for long enough periods to develop its characteristic blueish-grey soil colour due to anaerobic conditions. The soil classification in the GA and therefore its suitability for various purposes help in land use planning, and prioritizing their uses accordingly for production and conservation.

In the study, GA land management from 1987 to 2010 was evaluated and a projection for 2035 land use was made by employing various models such as the Markov Chain, multicriteria analysis and CASA (Carnegie Ames Stanford Approach) (Logofet and Lesnaya, 2000; Dönmez et al., 2013). The land use-related carbon zones (CAZ), the NPP and land cover served as the main criteria to assess a future land management model developed with global vegetation mapping. GA's vegetation varies by elevation. Red pine (*Pinus brucia*) is commonly seen from sea level to 700 m, as well as maquis (*Quercus*) which dominates the range between 900–1,200 m. And from 1,200–2,100 m, abies (*Abies cilicica*), juniper (*Juniperus*), black pine (*Pinus nigra*) and cedrus (*Cedrus libani*) flourish. The coastal zone is rich in halophytes and sand dunes grasses. The area above the timberline (2,100 m) is dominated by steppe vegetation. The natural distribution of the trees may be useful when choosing the right tree for the correct elevation for biomass production.

Organic carbon analysis

As part of the overall study, thematic maps of the GA were produced with various data sets including 1:25,000 scaled topographic, soil, and vegetation maps and remotely sensed images. Soil sampling locations were selected after determining the normalized difference vegetation index (NDVI) and net primary production with soil, slope and land use maps. In total, 421 soil samples were collected for total organic carbon determination.

To assess the GA land use, total organic carbon (TOC) and nitrogen (N) in the soil were measured. These properties are important to understand the impact of management patterns on land quality and thereby adopt future land use planning (Zhang et al., 2013). Soil carbon was analysed with a modified Walkley-Black method to adjust the incomplete digestion of the organic matter (Allison, 1960). For samples with more than 6 per cent organic matter, loss on ignition at 450 °C was applied to sieved and air-dried samples.

The soil organic carbon (SOC) was calculated as follows:

$$SOC_S = SOC_C\ (\%) \times BD\ (Mg\ m^{-3}) \times A(ha) \times LD\ (m) * 10^{-2}$$

Where: SOC_S: total amount of soil organic carbon for the given depth (t ha^{-1})
SOC_C: soil organic carbon content for the given depth (%)
BD: bulk density (Mg m^{-3})
A: area (ha)
LD: depth of soil layer (m)

Following the soil carbon analyses, humic and fulvic acid analyses for total nitrogen were performed (Lamar and Talbot, 2009), using samples with minimum, moderate and maximum carbon contents under various land uses to assess the soil carbon quality (Akça et al., 2010). The total soil nitrogen was estimated using the micro Kjeldahl method (Bremner and Keeney, 1965).

The Göksu anthroscape's facts and figures for bioeconomy assessment

The NPP was used to assess the anthroscape's bioeconomy potential in Göksu. As stated by Eswaran et al. (2010), NPP is the output of photosynthesis and respiration and is significant in global and regional carbon budgets, and has a function of the standing biomass along with soil organic carbon and is an indicator of healthy soil that supports high biomass production.

Land use

The GA offers various soil and climate types that are suitable for diverse land use, from citrus on alluvial plains to grazing small ruminants above the timberline. Habitation has generally been concentrated in the deep valley bottoms due to agricultural activities on alluvial soils, i.e. Fluvisols. The second settled area is relatively flat plateaus mostly used for cereal cultivation and raising livestock. Thus, the pressure from the increasing population changed the land use in GA from forests to agricultural purposes, which increased from around 229,436 ha in 1987 to 390,705 ha in 2010. The main driver of this change was agriculture, which offers higher income opportunities for locals than forests. Forested areas in Turkey are government entities, and, by law, locals cannot individually benefit from wood products as a source of income. The model revealed that by 2035, the GA's main land use would be agriculture, followed by forests, 49,736 ha of which have disappeared over the past 23 years. The projected

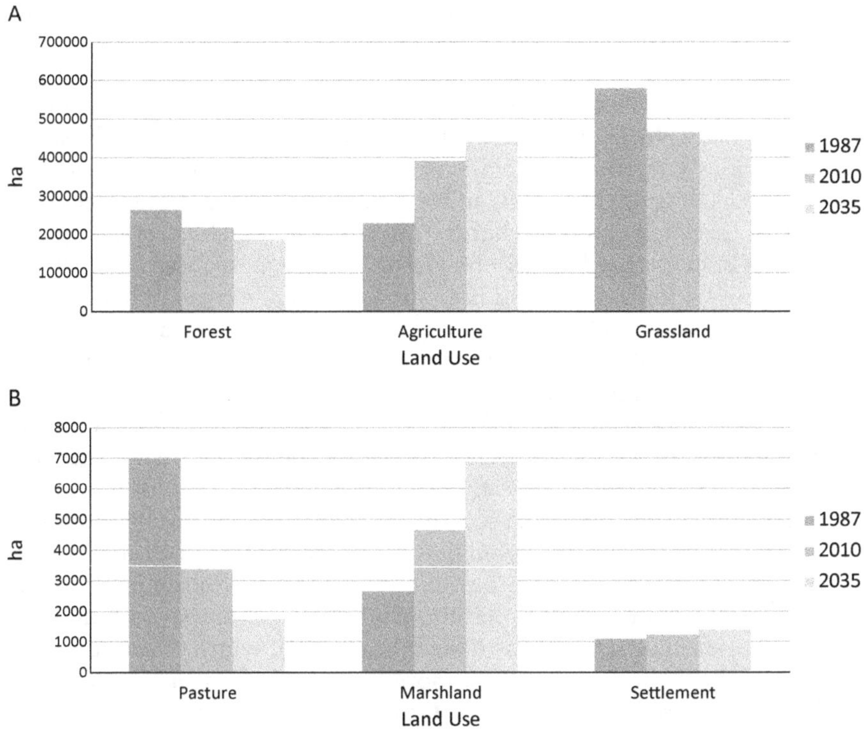

Figure 8.3 The land use trend in GA for forest, agriculture and grassland (A), and pasture, marshland and settlement (B), from 1987 to 2035 (hectares).

forest loss is 51,017 ha by 2035, which will have an adverse effect on the region's biomass resources, such as timber production. Grasslands, a mixture of grass land with rocky-stony areas with no or very sparse vegetation in the GA, will be decreasing due to the shift to agriculture. Despite the decrease, grasslands will still be the third largest land use category in the GA in 2035 (Figure 8.3 A). With serious reclamation measures and investments to protect grasslands in the GA, their potential for the production of forages can be realized. The projection reveals that the region's annual biomass output may increase, as agriculture will be the main driver of the economy. The high future biomass production, particularly, crop residues after harvest, could be used for bioenergy production, which in turn can enhance the bioeconomy potential of the GA.

Soil organic carbon

Bioeconomy-related activities in the GA focus mainly on food and fibre production rather than bioenergy, making soil carbon the possible main indicator for assessing the land use effects on GA's soil biomass production. The soil analyses revealed that the total organic carbon (TOC) in the GA varied from as low as 0.023 per cent in hazelnut orchards to as high as 6.34 per cent in natural vegetation of juniper-oak canopy, thus giving an average of 2.76 per cent for all the GA. The lowest TOC value for natural vegetation was found in an astragalus field at 0.17 per cent which shows uncontrolled livestock activities seriously threaten the biomass production of shallow and sloping soils in the GA. The highest TOC in the cultivated area was 6.04 per cent, observed in irrigated apple orchards. The lowest (hazelnut) and highest (apple) TOCs in orchard fields showed that carbon sequestration could succeed or fail with the same land use. The GA's TOC revealed that alluvial soils with intense agricultural uses have the lowest carbon accumulation, amounting to 2,899.6 gCm^{-2}. This low figure rose to as high as 28,529.2 gCm^{-2} in forested areas which may indicate the presence of low carbon-efficient agricultural activities in the GA, such as lack of legume rotation and removal of harvest leftovers from soils (Figure 8.4). The regeneration of degraded forests in the GA takes a long time, which means it is better to consider forests on the site as security zones to reduce runoff water and erosion. However, logging is ongoing for pine cone and oak acorn. The low soil TOC in agricultural areas is a common phenomenon in semi-arid land due to excess use of nitrogen fertilizers, insufficient crop rotation, low organic material input and excess tillage (Lal, 2013), which is also the case for the GA under agriculture. Thus, more leguminous fodder crop cultivation should be encouraged instead of cereals, in extensive agricultural areas. Priority should be given to agroforestry in areas with shallow soils on steep

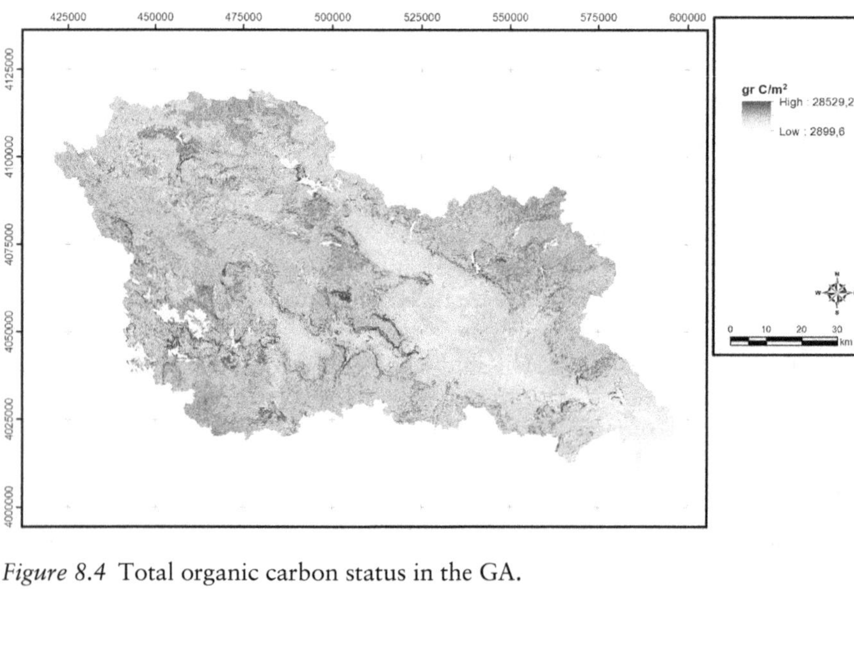

Figure 8.4 Total organic carbon status in the GA.

slopes which can be rented to locals for long periods. In the GA, grasslands need special attention, although the current biomass production is low. As grasslands take up high proportions of GA, they may be reclaimed for more biomass production by encouraging local annual plant species by seeding, which could also lead to decreasing water runoff.

Net primary production

Another factor affecting soil organic carbon is the NPP (Todd-Brown et al., 2014), particularly in the forested areas. The NPP in the GA was calculated using the CASA (Carnegie–Ames–Stanford Approach) model, which uses data sets of NDVI, climate parameters (temperature, precipitation, solar radiation), tree closure and soil-and-land cover. The NPP in the GA ranged widely from 0–1,856 gCm^{-2} (Figure 8.5). The highest NPPs for all vegetation cover in the GA were observed in May and June when temperatures and solar radiation were at their optimal values. Details of the GA's NPP revealed that agricultural activities, particularly in the irrigated areas of the GA, have a high NPP due to excess fertilizer and water use, followed by that of forests. However, this biomass production's input to soil carbon is lower than forests as all biomass is removed after the harvest and the high nitrogen fertilizer immediately decomposes the plant leftovers in the soil.

Current terrestrial carbon stocks and bioeconomy domains (BiD)

The soil carbon map as well as the soil texture and soil bulk density were used to calculate the soil organic carbon of the GA, with a soil organic

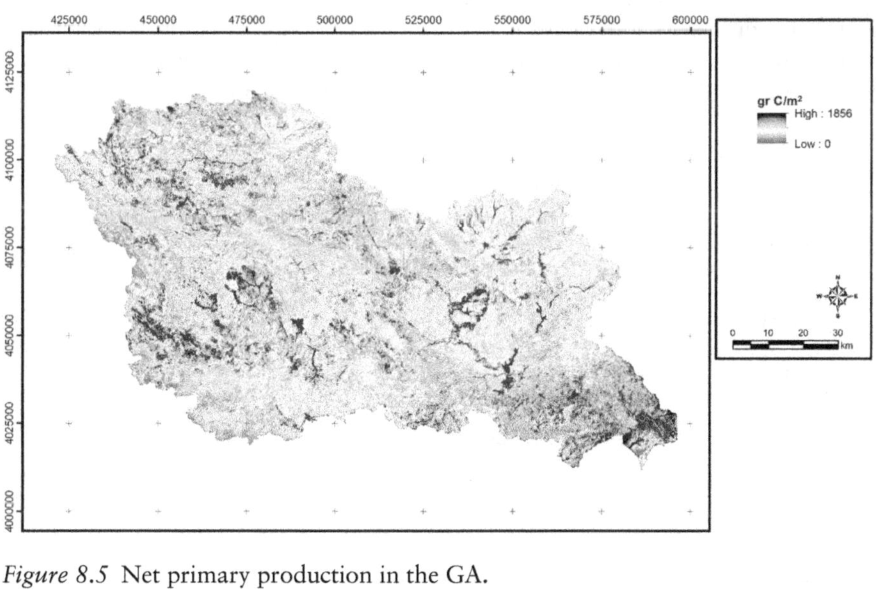

Figure 8.5 Net primary production in the GA.

carbon computation formula with a 100-pixel resolution. The TOC of soil from natural and agricultural vegetation showed significant variations with the cedar being the highest carbon-sequestrating tree with 11,959.7 gCm^{-2} while grassland exceeded agriculture's 8,375.8 gCm^{-2} due to dense grass vegetation (see Figure 8.5).

The dense cedar cover's high organic soil TOC may most likely be due to the low temperature and relatively high soil moisture which inhibit the decomposition of organic matter (Wang et al., 2017). Also, population pressure on forests in these areas is very low due to strict forest conservation regulations by the government. The high carbon value for oak revealed that water harvesting is an essential input to biomass production for sustainable bioeconomical domain (BiD) management in such a semi-arid anthroscape with proper tree species demanding low levels of water. The higher soil organic carbon of grasslands compared to the agricultural areas confirms the native plants' efficiency in carbon sequestration even without any management plan. The low values of juniper and black pines, compared to fir, cedar and oak, come from their sparse distribution in forests because these forests are close to settlements and grazing areas and locals have been using these trees for fuel and timber for centuries (Table 8.1).

The carbon accumulation zones (CAZ) were produced to assess the effects of landscape, i.e. the features of plants, topography, soil and geology affecting the soil TOC in the GA. The highest TOC-bearing CAZ was 20,556.4 gCm^{-2}, from a calcareous sandstone/siltstone/limestone-compound area comprising more than 30 per cent slope at 1,000–1,500 m. The texture of this CAZ is 30 per cent sand, 35 per cent silt and 35 per cent clay and the soil is 2 m deep. Crown closure of vegetation varied from 0–30 per cent. The CAZ with the lowest TOC of 3,537.27 gCm^{-2} was found on conglomerate-mudstones at 1,500–2,000 m on 10–30 per cent slopes in a mainly non-forested area with 0–30 per cent crown closure. The low carbon accumulation in this CAZ is most likely attributed to shallow soils

Table 8.1 Vegetation cover and mean amount of carbon stock in soil in the GA

Vegetation cover	*Mean carbon stock in soil, gCm^{-2}*
Juniper	9,925
Fir	10,270
Black pine	8,939
Red pine	8,323
Oak	11,024
Cedar	11,960
Agriculture	8,376
Grassland	9,058
Others	9,350

with little ability to support vegetation growth as well as the low water infiltration rate (Dehotin et al., 2015). According to Porter (1990), the GA's soils and climate are thus bottlenecks to develop biomass-oriented production due to shallow soils and low precipitation with high erosion rates.

The quality of TOC in the GA

Bioeconomy-related activities require healthy soil to support a high NPP which could in turn be used for bioenergy (Priefer et al., 2017). The soils' carbon stock by volume may not be enough to assess the soil quality. Carbon turnover rate is thus one of the tools used to evaluate soil quality (Guimarães et al., 2013). To assess the soil TOC turnover status, the active and passive organic carbon is measured in selected soils with low (<2 per cent), moderate (2–5 per cent) and high (>6 per cent) carbon content (Horwarth, 2015) (Table 8.2). Results revealed that high humic acid (HA)

Table 8.2 The organic carbon stability in selected soils of various land covers in the GA

Class	*Sample*	*Land use and age (years)*	*Organic carbon (%) (OC)*	*Humic acid (HA)*	*OC/HA*
Soils with high C (>6%C)	47	Apple 12	6.00	0.44	13.64
	64	Oak 30	6.76	0.56	12.07
	259	Juniper 50	6.76	0.51	13.25
	309	Black pine 50	6.49	0.52	12.48
	325	Cedar 50	5.94	0.50	11.88
	417	Red pine 40	6.68	0.50	13.36
Soils with moderate organic carbon (2-4%)	15	Oak 20	3.53	0.28	12.61
	66	Cedar 30	3.18	0.26	12.23
	113	Juniper 30	3.59	0.29	12.38
	114	Apple 10	2.35	0.19	12.37
	137	Red pine 30	3.71	0.30	12.37
	214	Cereal annual	2.06	0.16	12.88
	236	Olive 30	2.76	0.22	12.55
	265	Black pine 25	3.12	0.26	12.00
	371	Apple-sour cherry 15	3.21	0.25	12.84
Soils with low organic carbon (<2%)	50	Black pine 20	0.35	0.03	11.67
	51	Oak 20	0.53	0.04	13.25
	123	Red pine 20	0.76	0.06	12.67
	192	Olive 30	0.29	0.02	14.50
	258	Cedar 30	0.41	0.03	13.67
	343	Cereal annual	0.35	0.03	11.67
	390	Juniper 30	0.5	0.04	12.50
	414	Apple 5	0.32	0.03	10.67

is mainly found in long-standing vegetation cover, such as oak and juniper, while the agricultural uses show relatively low HA. HA has a linear relationship with the age of the plant cover. Forests and orchard plantations will thus have a positive effect on GA's TOCs if maintained for more than 40 years. We suggest using soil humic acid content to assess the soils' biomass-supporting ability as soils with a high HA most likely can support high NPP by stimulating root growth, nutrient uptake and crops' resistance to abiotic stresses (Canellas et al., 2015) that may boost BiDs' economic value. The efforts to increase soils' HA level will also need to incorporate more organic substances in the soil, such as pruning and harvest residues that will support both the soil's physical and chemical properties, such as water-holding capacity, aeration and microorganism biodiversity.

Future of soil carbon in the Göksu anthroscape

By 2035, based on the Markov Chain model, it is projected that the GA will have six major land use types; namely, forests, pastures, agriculture, grasslands, marshlands and settlements. To estimate the GA's carbon stocks, current terrestrial carbon stock data and 2035 land cover projection were integrated for a zonal statistical analysis. After classification, projection conversion and equating the data resolution for zonal statistical analysis (van Asselen and Seijmonsbergen, 2006), the trend and amount of TOC for the GA were produced (Table 8.3).

The projection revealed that TOC will be highest in grasslands and forest cover with $9{,}409.4\,gCm^{-2}$ and $9{,}398.9\,gCm^{-2}$ respectively. However, while grasslands will present the highest TOC, the increase in the GA will be 0.61 per cent until 2035, but in low TOC marshlands, the TOC sequestration will increase to 4.19 per cent by 2035. This may be due to the higher water supply to wetlands from upstream GA, along with the fact that the mild climate on the coast enables almost nine months of vegetative growth. The high potential of wetland biomass production can be used for GA's biofuel production where the regeneration rate of reeds is high and suggests that the coastal vegetation should be protected for its high biomass production.

Table 8.3 TOC trend in the GA from 2010 to 2035

Type of cover	*2010*	*2035*
Forest	9,274	9,399
Pasture	9,387	9,409
Agriculture	8,447	8,678
Grassland	8,967	9,383
Marshland	5,890	8,371
Settlements	7,966	710

Discussion

The Göksu anthroscape located in Mediterranean region of Turkey represents a wide range of integrated sustainable land use that humans have managed for millennia. The average total organic carbon accumulation, one of the main environmental management indicators of the Göksu anthroscape, which is a mixture of natural and agricultural landscape, is close to the 9,700 gCm^{-2} of soil TOC in the natural vegetation areas of the Mediterranean determined by Fernández-Romero et al. (2014). The lowest 0.023 per cent TOC, found in hazelnut orchards, to the highest 6.34 per cent, found in apple orchards, set the wide range of carbon sequestration of GA agricultural soils, which can be used as a base indicator for future land use planning. The forests, which already have one of the higher TOCs, will have an increasing trend in the GA, though the GA has limited topographic conditions (more than 6 per cent slope) and soil conditions (shallow, ranging from 0–60 m and eroded). Sufficient forest and agriculture management will support the population but the GA's present land use may seem inadequate for biomass production for bioenergy. A study is needed to calculate the agricultural and forestry NPP and the residue potential of the GA, which is most probably high, as the soil organic carbon content is high under forest covers and in certain agricultural crops such as apples. The soil organic carbon of grasslands with rocky-stony areas, even when no management plan is applied, appears to be higher than that of agricultural lands, which reveals the potential of native plant species to produce high biomass. These areas require interventions by the government to undertake reclamation as the investment costs for terracing and seeding are beyond the capacity of local budgets. Reclaiming grasslands can create new grazing areas for small ruminants, which will decrease the grazing intensity of the pastures and at the same time increase farmers' income.

Similarly, afforestation should be undertaken by considering the elevation of the GA, because it is one of the major determinants of natural vegetation. For example, cedar has a higher carbon sequestration efficiency than red pine due to high altitudes, high water retention capacity and the strict forest conservation measures of the government. The government deems it necessary to keep forest cover with species specific to each elevation as the main land use in the GA highlands (Kirby and Potvin, 2007), which may be used for biomass production. The pressure from the increasing population to shift to agriculture may reduce the soil organic carbon by 20–50 per cent (Murty et al., 2002; Álvaro-Fuentes et al., 2008). Thus, irrigated apple and sour-cherry orchards are suggested for high soil carbon sequestration in the GA at altitudes above 1,000 m if or when land is allocated for agriculture. Kapur et al. (2019) determined that the Pleistocene and Holocene terraces of south and east Turkey are high NPP areas, managed as agro-ecosystems and allocated for indigenous crops grown on the calcretes (caliches) on the terraces and the limestones of the karst. Pastures in the GA will contain high TOC stocks. However, the traditionally grazed livestock habits may threaten these areas' high TOC potential. Grazing needs thus need to be rotated in the GA by shifting to grasslands at high altitudes, or grazing may need to be avoided in vulnerable CAZs, such as in the highlands and grasslands of the GA (Wen et al., 2016).

The HA is another indicator of organic matter maturation (stabilization) in soils (Akça et al., 2010). The GA's high HA contents are mainly found in areas with forest cover aged more than 40 years and in irrigated orchards. This may suggest that forest cover be kept for more than 50 years before logging. The longer lifespan of forests will enhance the soils' ecosystem functions by retaining more water, increasing the soil carbon and providing nutrients to macro- and microorganisms. The enhanced ecosystem services such as pollination will support orchards, namely, olives, apples, cherries, or vineyards in the GA. Orchards will provide more wood during their economic production period for landowners. This wood sources in turn can be used for bioenergy production.

The outcomes of the study have revealed that coastal marshlands will have a high potential for biomass production due to sufficient water supply from the highlands, and the mild climate that supports around nine months of vegetative growth . Future projections are required to plan land uses of any BiD as their carrying capacity will be affected by climatic and population pressure.

The Göksu anthroscape's organic carbon will increase slightly until 2035, representing a calcrete anthroscape showcasing the human ability of challenging harsh conditions via their environmentally friendly experiences amassed for millennia. The organic carbon is high not only in natural forests but also in grasslands and agricultural lands. Thus, a combination of forest-agriculture-pasture will secure an income for the locals in the GA without sacrificing soil organic carbon that supports biomass production. In summary, the Göksu anthroscape will seed BiD with integrated land management that may prevent the formation of 'a hundred dead cities' as in ancient times, while securing income for the local population through developing sustainable bio-resource based economy.

Conclusion

Excluding humans from environmental management has mainly failed, and environmental management is not possible without involving local communities and their interests. The human-shaped landscape, the GA in this case, was quantified by present and future NPP and TOC, which are the outcomes of climate, topography vegetation and human interference. Assessing the NPP alone for biomass production is not sufficient to measure land-use sustainability. For example, a high irrigation rate increases NPP, but if all the biomass is removed after harvest, the soil organic carbon cannot benefit from high production. Therefore, soil organic carbon dynamics should be incorporated into the evaluation of land-use sustainability. The GA has shown sustainable examples of calcrete-dominated land-use in the Mediterranean region. This study sought to quantify the carbon dynamics of the anthroscape to suggest how to establish a new approach to use carbon accumulations as an indicator to develop the regions. Preserving these natural and man-made surfaces with enhanced soil quality will improve crop productivity and protect the land from erosion and degradation which in turn will yield large amounts of biomass. Moreover, the crop/soil management practices in these domains, based biomass production, will enhance CO_2 sequestration. This projected high biomass production is crucial to reduce Turkey's dependence on imported fossil fuels.

The anthroscape-based BiD approach is expected to connect the effects of the land use transition to their socio-economic or political drivers. Socio-economic analyses such as that of crop patterns (Umetsu et al., 2007) should be undertaken to evaluate whether biomass production-related activities in the GA fall within the carrying capacity of the natural resources.

The BiD approach will be a powerful quantitative tool for local communities/administrations and relevant groups to reach future land and water use decisions with a bottom-up approach. The final TOC quantification of an anthroscape will reveal the magnitude and the distribution of biomass productivity on a regional scale. This will help designate appropriate sites for ecosystem management while satisfying the human need for economic survival.

The chapter also shows the potential of various indicators, including NPP and the soil organic carbon, in future land-use planning, and the need to measure such indicators using robust scientific methods. This will help planners to become aware of the production potential and the sustainability of the region for better land-use planning. Any future bioeconomy development initiatives will need a good database of sustainability indicators and mapping of the regions, as evident from the findings of the study in the GA.

References

Akça, E., Bal, Y., Kapur, S. and Eswaran, H. (2009) 'Human-induced late Holocene degradation of the Asi (Orontes) delta, Samandağ, Southern Turkey', in *Land Degradation and Rehabilitation: Dryland Ecosystems*. Papers presented at the Fourth International Conference on Land Degradation, Cartagena, Murcia, Spain, 12–17 September 2004, pp. 17–26.

Akça, E., Çimrin, K.M., Ryan, J., Nagano, T., Topaksu, M. and Kapur, S. (2008) 'Differentiating the natural and man-made terraces of Lake Van, Eastern Anatolia, utilizing earth science methods', *Lakes & Reservoirs: Research & Management*, vol. 13, no. 1, pp. 83–93.

Akça, E., Kapur, S., Tanaka, Y., Kaya, Z., Bedestenci, H.C. and Yakti, S. (2010) 'Afforestation effect on soil quality of sand dunes', *Polish Journal of Environmental Studies*, vol. 19, no. 6, pp. 1109–1116.

Allison, L.E. (1960) 'Wet-combustion apparatus and procedure for organic and inorganic carbon in soil', *Soil Science Society of America Journal*, vol. 24, no. 1, pp. 36–40.

Álvaro-Fuentes, J., López, M.V., Cantero-Martínez, C., and Arrúe, J.L. (2008) 'Tillage effects on soil organic carbon fractions in Mediterranean dryland agroecosystems', *Soil Science Society of America Journal*, vol. 72, no, 2, pp. 541–547.

Başaran, M. (2003) 'Carbon stocks of Turkey', *Journal of Faculty of Agriculture*, vol. 8, pp. 31–36.

Bremner, J.M., Keeney, D.R. (1965) 'Steam distillation methods for determination ammonium, nitrate and nitrite', *Analytica Chimica Acta*, vol. 32, pp. 485–495.

Canellas, L.P., Olivares, F.L., Aguiar, N.O., Jones, D.L., Nebbioso, A., Mazzei, P. and Piccolo, A. (2015) 'Humic and fulvic acids as biostimulants in horticulture', *Scientia Horticulturae*, vol. 196, pp. 15–27.

Chasek, P., Safriel, U., Shikongo, S. and Fuhrman, V.F. (2015) 'Operationalizing zero net land degradation: The next stage in international efforts to combat desertification?', *Journal of Arid Environments*, vol. 112, pp. 5–13.

Dehotin, J., Breil, P., Braud, I., de Lavenne, A., Lagouy, M. and Sarrazin, B. (2015) 'Detecting surface runoff location in a small catchment using distributed and simple observation method', *Journal of Hydrology*, vol. 525, pp. 113–129.

Dönmez, C., Berberoğlu, S., Forrest, M., Çilek, A. and Hickler, T. (2013) 'Comparing process-based net primary productivity models in a Mediterranean watershed', *International Archives of the Photogrammetry, Remote Sensing and Spatial Information Sciences*, vol. XL-7/W2, p. 67.

Erşahin, S., Kapur, S., Aydın, G., Akça, E., Tolunay, D., Görücü, Ö., Karahan, G. et al. (2017) 'Terrestrial ecosystem carbon dynamics as influenced by land use and climate: Carbon management, technologies, and trends in Mediterranean ecosystems', *Science*, vol. 15, pp. 27–44.

Eswaran, H., Berberoğlu, S., Cangir, C., Boyraz, D., Zucca, C., Özevren, E. and Çelik, I. (2010) 'The anthroscape approach in sustainable land use', in S. Kapur, H. Eswaran and W.E.H. Blum (eds) *Sustainable Land Management*. Berlin: Springer, pp. 1–50.

Eswaran, H., Kapur, S., Akça, E., Reich, P., Mahmoodi, S. and Vearasilp, T. (2005) 'Anthroscapes: A landscape unit for assessment of human impact on land systems', in J.E. Yang, T.M. Sa and J.J. Kim (eds) *Application of the Emerging Soil Research to the Conservation of Agricultural Ecosystems*. Seoul: The Korean Society of Soil Science and Fertilizers, pp. 175–192.

Fernández-Romero, M.L., Lozano-García, B. and Parras-Alcántara, L. (2014) 'Topography and land use change effects on the soil organic carbon stock of forest soils in Mediterranean natural areas', *Agriculture, Ecosystems & Environment*, vol. 195, pp. 1–9.

Fujinawa, K. (2010) 'Anthroscape of the Mediterranean coastal area in the context of hydrogeology: Projected impacts of climate change', in S. Kapur, H. Eswaran and W.E.H. Blum (eds) *Sustainable Land Management*. Berlin: Springer, pp. 311–332.

Guenet, B., Camino-Serrano, M., Ciais, P., Tifafi, M., Maignan, F., Soong, J.L. and Janssens, I.A. (2018) 'Impact of priming on global soil carbon stocks', *Global Change Biology*, vol. 24, no. 5, pp. 1873–1883.

Guimarães, D.V., Gonzaga, M.I.S., da Silva, T.O., da Silva, T.L., da Silva Dias, N. and Matias, M.I.S. (2013) 'Soil organic matter pools and carbon fractions in soil under different land uses', *Soil and Tillage Research*, vol. 126, pp. 177–182.

Horwarth, W. (2015) 'Carbon cycling: The dynamics and formation of organic matter', *Soil Microbiology, Ecology and Biochemistry*, vol. 4, pp. 339–382.

Jacobsen, T. and Adams, R.M. (1958) 'Salt and silt in ancient Mesopotamian agriculture', *Science*, vol. 128, no. 3334, pp. 1251–1258.

Kapur, S., Akça, E., Zucca, C., Berberoğlu, S. and Miavaghi, S.R. (2019) 'Anthroscapes: A robust basis for mapping land quality and sustainable land use patterns', in F Yenişehirlioğlu and E Özveren (eds) *Eastern Mediterranean Port Cities*. Cham: Springer, pp. 63–77.

Kirby, K.R. and Potvin, C. (2007) 'Variation in carbon storage among tree species: implications for the management of a small-scale carbon sink project', *Forest Ecology and Management*, vol. 246, no. 2, pp. 208–221.

Lal, R. (2013) 'Intensive agriculture and the soil carbon pool', *Journal of Crop Improvement*, vol. 27, no. 6, pp. 735–751.

Lamar R.T. and Talbot K.H. (2009) 'Critical comparison of humid acid test methods', *Communication in Soil Science and Plant Analysis*, vol. 40, pp. 2309–2322.

Lev-Yadun, S., Gopher, A. and Abbo, S. (2000) 'The cradle of agriculture', *Science*, vol. 288, no. 5471, pp. 1602–1603.

Logofet, D.O. and Lesnaya, E.V. (2000) 'The mathematics of Markov models: What Markov chains can really predict in forest successions', *Ecological Modeling*, vol. 126, pp. 285–298.

Lowdermilk, W.C. (1950) *Conquest of the Land through Seven Thousand Years*. Washington, DC: US Government Printing Office.

Malhi, Y., Doughty, C. and Galbraith, D. (2011) 'The allocation of ecosystem net primary productivity in tropical forests', *Philosophical Transactions of the Royal Society B, Biological Sciences*, vol. 366, pp. 3225–3245.

Mattern, J. (1944) *A travers les villes mortes de Haute Syrie*. Beyrouth: Imprimerie Catholique.

Murphy, D.J. (2007) *People, Plants and Genes: The Story of Crops and Humanity*. Oxford: Oxford University Press.

Murty, D., Kirschbaum, M.U.F., McMurtrie, R.E. and McGilvray, H. (2002) 'Does conversion of forest to agricultural land change soil carbon and nitrogen? A review of the literature', *Global Change Biology*, vol. 8, pp. 105–123.

Porter, M. (1990) *The Competitive Advantage of Nations*. New York: The Free Press.

Priefer, C., Jörissen, J. and Frör, O. (2017) 'Pathways to shape the bioeconomy', *Resources*, vol. 6, no. 1, pp. 1–24.

Rasmussen, R.K. (2010) *Agriculture in History*. Hackensack, NJ: Salem Press.

Şahin, Ü. (2018) 'Carbon lock-in in Turkey: A comparative perspective of low-carbon transition with Germany and Poland'. İstanbul: İstanbul Policy Center.

Şengül, E. (2017) 'Turkey's energy import bill up by 37% in 2017'. Available at: www.aa.com.tr/en/energy/finance/turkeys-energy-import-bill-up-by-37-in-2017/18644

Todd-Brown, K.E.O., Randerson, J.T., Hopkins, F., Arora, V., Hajima, T., Jones, C., Shevliakova, E., et al. (2014) 'Changes in soil organic carbon storage predicted by Earth system models during the 21st century', *Biogeosciences*, vol. 11, pp. 2341–2356.

Umetsu, C., Palanisami, K., Coşkun, Z., Donma, S., Nagano, T., Fujihara, Y. and Tanaka, K. (2007) 'Climate change and alternative cropping patterns in Lower Seyhan Irrigation Project: A regional simulation analysis with MRI-GCM and CSSR-GCM'. Available at; https://ecomod.net/sites/default/files/document-conference/ecomod2007-rum/52.pdf

van Asselen, S. and Seijmonsbergen, A.C. (2006) 'Expert-driven semi-automated geomorphological mapping for a mountainous area using a laser DTM', *Geomorphology*, vol. 78, no. 3, pp. 309–320.

Wang, T., Kang, F., Cheng, X., Han, H., Bai, Y. and Ma, J. (2017) 'Spatial variability of organic carbon and total nitrogen in the soils of a subalpine forested catchment at Mt. Taiyue, China', *CATENA*, vol. 155, pp. 41–52.

Wen, D., He, N. and Zhang, J. (2016) 'Dynamics of soil organic carbon and aggregate stability with grazing exclusion in the Inner Mongolian grasslands'. Available at: https://doi.org/10.1371/journal.pone.0146757

Zhang, C., Liu, G., Xue, S. and Sun, C. (2013) 'Soil organic carbon and total nitrogen storage as affected by land use in a small watershed of the Loess Plateau, China', *European Journal of Soil Biology*, vol. 54, pp. 16–24.

9 Centrum Woods Vision

Creating new value chains for sustainable woods management in Japan

Takanori Nagano, Yuka Bando, Yutaka Okano Daisuke Maki and Takahiro Ueyama

Introduction

After the Second World War, Japan gradually built up the nation's wealth by an emphasis on industry, trade and processing imported natural resources into end products. As Japan's land area is mainly composed of mountains (75 per cent), industrial bases were developed mainly in the plains and around port cities facing the Pacific Ocean. Industrial value chains were created for imported goods such as oil and other raw materials, and for export goods such as cars and electronic items. In 2018, Japan ranked third in gross domestic product (GDP) globally, only after the United States of America and China.

Japan's ecological footprint for consumption was 4.1 global hectares (gha) in 2006, significantly higher than the world's average of 2.6 gha. The ecological footprint, by definition, is the amount of productive land and sea area required to produce the resources a person consumes and to absorb his waste, including CO_2 emissions in comparison to the biocapacity, and the ability of ecosystems to regenerate resources. In other words, if everyone on the Earth pursued a lifestyle like a Japanese citizen, we might need 2.3 times the size of the Earth (4.1 gha divided by the average biologically productive area per person worldwide of 1.8 gha). A high ecological footprint is the result of the high carbon footprint of Japan, which constitutes as much as 65 per cent of the total ecological footprint (WWF Japan, 2010). In 2016, Japan was only 6.0 per cent self-sufficient in primary energy and much of the energy demand was met by imported fossil fuels. This is not sustainable, given the current ecological footprint of Japan.

Geographically, Japan has great potential to exploit its own natural resources, especially the forests and its coasts through following a sustainable bioeconomy path. It is an island situated in the Far East with a territory of 378,000 km^2. Japan is comprised of 6,852 islands, stretching over the length of 3,000 km from 24°N to 46° N, ranging from various climatic zones between sub-tropical and sub-Arctic. The country receives an average annual precipitation of 1,718 mm which is nearly twice as much as the world average precipitation of 880 mm. Forests or woods cover nearly 67 per cent of its territory that amount to 4.9 billion m^3 of wood resources. It has a long coastline of about 33,889 km and an exclusive economic zone covering 4,470,000 km^2 which is the eighth largest in the world. Until the 1960s, the main energy sources for Japanese households were charcoal and firewood. Secondary forests around villages and urban areas were intensively exploited to meet the domestic energy demand, in addition to gathering plant leaves and branches to fertilize farmlands. Seafood and salt harvested in the coastal areas were used for human consumption and to fertilize the land. Apart from privately owned land, there were vast commons which were managed by local communities. This traditional land use, which continued for centuries, created distinctive ecosystems that were called 'Saotyama' and 'Satoumi'. These are defined as dynamic mosaics of socio-ecological systems, which produce a bundle of ecological services for human well-being (Duraiappah et al., 2012).

According to the recent integrated assessment of the biodiversity and ecosystem service report, issued by the Ministry of Environment in 2016, there has been a large decline in ecosystem supporting services in the past 50 years in Japan. Some of the agricultural, marine and forest resource supply has decreased to less than 50 per cent of the production peaks. Marine support services have drastically reduced, mainly due to overfishing and destruction of habitats. Fish catches of coastal fishery in 2016 was only 99,000 tons, which was roughly half of the catch in the late 1960s (Fisheries Agency, 2017).

In the agriculture sector, growth has weakened due to high international competition and smaller business scale, despite the modernization of the rural infrastructure, which was relatively well supported with abundant regional investment by the central government. Therefore, under-use of productive land due to the increase of imports occurred over the years. Currently, cereals, vegetables and vegetable products are only 45–60 per cent, 75 per cent and 40 per cent of the production peaks, respectively. According to a recent report, the calorie-based food self-supply rate in Japan is less than 40 per cent of the total demand (Ministry of Environment, 2017). This shows the high dependency on imported resources, despite the large potential that can be tapped from its own resources. In 2010, about 7.9 per cent of agricultural land was abandoned due to various structural and social changes happening in the rural landscape of Japan.

Japan's forestry sector has been sluggish since import liberalization in the late 1960s, and this led to wood or forest owners abandoning forest (e.g. thinning) maintenance and management. While 67 per cent of Japanese territory is under forests, a major part of this is dense plantation forests with low economic value. Due to low economic returns from agriculture and forestry sectors, out-migration from rural areas to cities increased in search of employment. This pattern is gradually expanding, affecting the social landscape in the rural areas of Japan at an unprecedented rate. Today, rural areas are dominated by an ageing and declining population, compared to urban areas. Due to globalization, small and medium-sized manufacturing units have been shifted to other countries since the end of the twentieth century, causing further decline in employment opportunities in rural areas.

Faced with a rapid population decline, the government promoted the merger of municipalities to reduce administrative costs. In 2014, a governmental expert committee announced the concept of 'municipality at risk of extinction'. This refers to municipalities where the female population between the ages of 20–39 is projected to decrease by more than 50 per cent from the numbers in 2010 by 2040. Nearly half of all the Japanese municipalities were in this category, highlighting the serious situation the country is facing (Japan Policy Council, 2014). In 2014, the government announced an initiative, based on a local revitalization project model, as a countermeasure to reverse the current demographic changes and migration trend (Cabinet Office, 2014). However, it is not an easy task to check or reverse the migration to urban areas, which has led to population concentrations in Tokyo and other major cities. Unless economic and livelihood opportunities are created and assured in rural areas, outmigration of youth to cities cannot be stopped. A country like Japan, therefore, needs a long-term strategy and vision to promote the rural economy with adequate investments in agriculture, forestry and other sectors.

This chapter briefly outlines the Japanese rural and urban landscape changes, the population dynamics, followed by a brief description of the new bioeconomy strategy. Nishiawakura's bioeconomy development model is then presented, demonstrating how a multi-faceted approach using forest and energy resources can help local communities find sustainable livelihood paths while reducing the ecological footprints.

Landscape and social changes in Japan

A brief history of forestry in Japan

Of Japan's total land area of 37.8 million hectares, forests occupy about two-thirds of the land area. Of these, 10.2 million hectares, or 40 per cent, are artificially planted forests. The forest composition is mainly constituted of 44 per cent cedar, 25 per cent cypress, 10 per cent larch and 21 per cent other species (Forestry Agency, 2019). In Japan, tree planting for timber was said to have started in the Yoshino region of west Japan, during the Muromachi Period (1336–1557). Later in the Edo Period (1603–1868), lumber production and afforestation spread in various areas of Japan (Nakashizuka and Kikuzawa, 2019). Subsequently, coniferous cedar and cypress were planted because of their good splitting properties. In the Meiji period (1868–1912), the newly established government designated the forest managed by the Tokugawa Shogunate as a national forest, which was 10 per cent the size of the entire land area of Japan and half the size of the royal forest. Subsequently, forest management was westernized, partly based on the German forestry model. Because of Japan's steeper landscape and heavier snowfall, natural regeneration was difficult, and afforestation became labour-intensive and expensive.

During the Second World War, woods in Japan were cut down and devastated due to a shortage of timber supplies. In addition, the demand for timber increased with the post-war reconstruction, and logging continued in the woods near settlements (Satoyama) and in natural forests more upstream (Okuyama), which were later converted to cedar and cypress plantations. This expansion of afforestation lasted from 1950 through to 1970. However, afforestation in Okuyama due to several challenges and soaring wages hampered the maintenance work. The liberalization of timber imports in the late 1960s led to a steep fall in the prices of domestic timber. Following this, forest owners gave up thinning and other necessary maintenance work. As such, domestic timber supply peaked at 70 million m^3 in 1957 and sharply decreased thereafter, hitting its historical low of 16.92 million m^3 in 2002. It then gradually increased to 29.66 million m^3 in 2017 due to changes in the forest management policy. Currently, only 36.2 per cent of the demand is met, leaving a large gap between the demand and supply, which clearly demonstrates the potential to exploit the large reserves of national wood resources (Forestry Agency, 2019).

The total biomass in Japanese woods is estimated to be about 5.2 billion m^3 as of 2017, with approximately 60 per cent of it (3.3 billion m^3) composed of planted woods. The total accumulation is increasing each year due to less logging. Overall, 57 per cent of the woods are privately owned, about 12 per cent are public, and the remaining 31 per cent are national forests. Private forests occupy 65 per cent of the planted forests (ibid.) and the major part of the planted forests are 50–60 years old.

Changing demographics and depopulation in Japan

In Japan, the population peaked at 128.08 million in 2008 and then entered a decline. It is further expected to decrease to 86.74 million by 2060. The decline is happening faster in rural areas, mainly caused by the out-migration of young people to cities. There were three peak periods in the past: (1) the first occurring in high economic growth period in the 1960–1970s; (2) the second was during the 1980–1990s when the Japanese economy flourished; and (3) the third period was after 2000, when the domestic economy suffered with the transfer of production bases to overseas. The concentration in the city of Tokyo and the suburbs is remarkable. The difference between the number of people moving into the Tokyo area and those moving out of the area, is proportional to the difference in effective job opening rates between Tokyo and other areas. In addition, the differences in the number of people moving into the three major metropolitan areas of Tokyo, Nagoya and Osaka to those moving out of these areas, are in proportion to the income gap (Masuda, 2014). In other words, the growing employment and income gap between urban and rural areas is the main driving force for migration. The rural areas are unable to promote population growth as their employment opportunities are decreasing due to declining public investment and production bases that are moving abroad. It is only the service sector, such as medical and nursing care, schools and local administration, that is currently supporting local employment. However, with the super-aging of urban areas, the need for medical and nursing care will grow in cities, which may result in further population outflow from rural areas or an increase in immigration.

At present, the senior population in urban areas is still growing but in rural areas it is plateauing or starting to decline, which means population decline will happen rapidly there. Population recovery is greatly influenced by the population of young females aged between 20–39 in an area. However, by 2040, it is estimated that the number of municipalities in Japan where the young female population will drop by more than 50 per cent (compared to 2010 figures) will be 896 out of 1,799 across the entire country. These are defined as 'municipalities at risk of extinction' (Masuda, 2014). The number of municipalities which will have a population of less than 10,000 in 2040 is estimated to be 523. The changing demographic trends will have serious socio-economic consequences on these municipalities and rural societies of Japan, unless drastic measures are taken to reverse or check the trends in the coming decades. Alternative paradigms such as bioeconomy development could offer promising opportunities to attract young people to move to rural areas. In countries such as Norway, rural reconstruction through bioeconomy initiatives is proving to be successful in keeping the population in rural areas. The Japanese government's rural policy enables the rural facilities to be maintained, including schools, health care and transport. However, in recent years the government has started to merge rural municipalities to cut down costs.

Mergers of municipalities and implications for rural development

Mergers of municipalities in Japan have seen three major developments in the past. The first was during the introduction of the municipality system in 1889, when 71,314 villages across Japan were merged into 15,859 municipalities. The second was when the town merger promotion law came into effect in 1953 where the main goal was to authorize only municipalities with more than 8,000 people to efficiently run junior high schools. The third was in 1956 when another law to promote the mergers was enacted to reduce the number of towns to one-third. By 1962, the number of towns and villages had dropped to 3,472. There was no change for the following 40 years until 2000, when the cabinet made the decision to reduce the number of municipalities to around 1,000. The main reasons for enforcing mergers were: (1) to improve the financial efficiency of the local administration; (2) to cope with the shrinking fiscal scale and growing medical welfare service, despite the declining birth rate and ageing society; (3) to accommodate expanding livelihood spheres; and (4) to strengthen the administrative and financial basis of municipalities towards decentralization (Ministry of Internal Affairs and Communications, 2010). As a result, the total numbers of basic municipalities were reduced to 1,799 in 2019. While the above goals were achieved to a certain extent, mergers in the process also had their disadvantages such as: (1) a fall in the vitality of the former municipalities in peripheral areas; (2) undue recognition of the voices of residents by the administration; (3) decline in services; and (4) loss of toponym and cultural diversity. Municipalities are responsible for the management of natural resources in their area. In that sense, mergers bring loss of knowledge about the local natural resources, which is disadvantageous for bioeconomy development.

Bioeconomy and other related strategies in Japan

In June 2019, the Cabinet Office of the Government of Japan announced its latest bioeconomy strategy. The future society described in this strategy paper would be built on a circular economy approach where wastes from one industry will be converted to resources for other industries in the vicinity. It will introduce data-driven management to primary industries for innovation and efficient management. Biotechnology in this context is expected to realize sustainable production with low environmental footprints (biofoundry). Health and medical care will be closely linked by data science and biotechnology to realize preventive approaches. High competitiveness in genome drug discovery and regenerative medicine will be pursued. Robotics and sensing technology will assist bio-processes. Advanced timber will replace concrete for building materials (Cabinet Office, 2019). Similar to the American and Chinese strategies, a fusion of data science, genome and biotechnology is addressed as a core technology in the strategy. Though technologically ambitious, there is a concern about a lack of notion about ecosystem services and rural society in the strategy, which is an integral part of sustainable bioeconomy development. As illustrated by the United Nations Sustainable Development Goals (SDGs), there are many human, societal and ecological dimensions that need to be addressed to realize sustainable development.

To promote both the circular economy and bioeconomy pathways, increasing the rate of renewable energy is important. The Japanese government announced its basic energy plan in July 2018 (Ministry of Economy, Industry and Commerce, 2018). It is aiming to increase the rate of renewable energy in power generation from 15 per cent (2016) to 22–24 per cent by 2030, which seems a very modest increase compared to the plans of other countries. By 2050, the government aims to accomplish a total decarbonization with renewable energy being the main energy source. However, details on the technical knowhow to be used to achieve the set goal are not clearly specified.

The potential for rural bioeconomy development in different sectors relevant for Japan

The following section provides some specific examples showing the potential for bioeconomy development in different sectors, including forests, agriculture and fisheries that are relevant to the rural areas of Japan located in the mountains and along the coastline.

As a follow-up to the bioeconomy strategy, forest or wood resources need to be more proactively used in Japan to reduce its carbon or ecological footprint. To maximize carbon absorption, thinning of abandoned forest needs to be carried out. Demand for domestic lumber declined due to the decrease in numbers of traditional houses and the fire regulations which prohibited use of flammable materials for exterior cladding in densely populated areas. New use of wood therefore is necessary. A good example is the cross-laminated timbers, which could be a promising alternative to replace concrete materials. However, there is a need to develop a viable technology that is affordable and environmentally friendly.

Similarly, in the agriculture sector, domestic food production needs to be increased with an emphasis on the value chain enhancement of selected food crops. There is excessive production of rice and too much dependence on imports for other crops, especially for cattle feed. In addition, agricultural law prohibited general enterprises to own or run farms until 2009. With the new policy, the enterprises can carry out farming on rented land. As of 2017, nearly 3,030 enterprises entered the farming business and contribute to the diversification of food production. Thus, the private sector can use some part of the abandoned agricultural areas. The area of abandoned agricultural fields amounts to 38,300 ha and more than half of those are small parcels situated in mountainous areas. Although fertile land and plenty of clean water are available, small farms remain abandoned due to low labour efficiency. New public-private partnerships and novel land uses are needed to activate the bioeconomy. One such example is the Yamaga Silk factory in Kumamoto, which runs an innovative silk farm factory. Because of its high productivity and quality, the domestic sericulture industry, which was the main source of income for mountainous communities in this region before the Second World War, and had almost became extinct has now been revived.

In the fisheries sector, there is great potential for improvements to make inshore fisheries more sustainable. The long coastline of Japan needs to be managed properly to restore the coastal environment combined with regulations for sustainable fishing and aquaculture. Coastal ecosystem services declined in Japan mainly due to the conversion of the natural coastal areas to industrial and urban areas. Construction of artificial islands by landfill adversely affected inshore ecosystem services (Duraiappah et al., 2012). Due to the sharp decline in fish catches inshore and offshore, the numbers of fisherman are sharply decreasing in Japan. Although new technologies like complete aquaculture for tuna and on-land aquaculture for some fish species are being introduced, total aquaculture production is decreasing. Fisheries rights are granted only to existing fishermen's union, a practice which until now has inhibited the participation of innovative enterprises. Fishery management in Japan is not good at introducing individual quotas or individual transferable quotas, which has become the norm among advanced fishing countries.

In a globalized economy there will always be serious competition between countries and regions to produce and market niche products. A long-term strategy supported by technological innovation and investments is required to pursue sustainable development pathways. Further, an innovative value chain needs to be created, using bio-resources and the bio-residues to lower the ecological footprint. Sustainable bioeconomy models must be developed locally that can be replicated in other regions of Japan. In the following section, a bioeconomy model comprising of several initiatives covering various sectors currently in operation in Nishiawakura village, located in Okayama Prefecture in Japan, will be discussed.

The case of Nishiawakura village: a 'Centrum Woods Vision' in Okayama Prefecture

Nishiawakura village, located in Okayama Prefecture in Japan, is a typical example where plantation forests are matured and under-utilized. The village community decided to revitalize the area through sustainable utilization of natural resources. In 2008, Nishiawakura declared it would follow the 'Centrum Woods Vision' initiative, to develop a new value chain of wood resources and increased use of renewable energy (Nishiawakura Village Council, 2019a). Young people from inside and outside the village supported the initiative and contributed to this development as entrepreneurs. Nishiawakura's pursuit of sustainability through multifaceted activities based on natural resources is at the forefront of the upcoming bioeconomy society in Japan. With confidence and based on the experience so far, we can claim that the initiative simultaneously pursues the following Sustainable Development Goals:

- Goal 5 (gender equality)
- Goal 7 (affordable and clean energy)
- Goal 9 (industry, innovation and infrastructure)
- Goal 10 (reduced inequality)

- Goal 12 (responsible consumption and production)
- Goal 13 (climate action)
- Goal 14 (life below water)
- Goal 15 (life on land).

Study area

Nishiawakura village is located in the Aida-gun, the north-eastern end of Okayama Prefecture, a mountainous area adjacent to Tottori Prefecture and Hyogo Prefecture. The village is located in a narrow valley of Yoshino River which runs southward to Setouchi Sea. The total area of the village is 57.97 km^2 with an altitude ranging from 263–1280 m. The average annual temperature and precipitation are 11 °C and 2,000 mm, respectively. It often snows during winter in the area. About 95 per cent of the area is woods, of which 84 per cent are artificial cedar and cypress woods (Nishiawakura Village Council, 2019b). The population is 1,455 as of June 2019, of which 33.1 per cent is aged over 65. Nishiawakura, just like any other mountainous village around, is experiencing depopulation and ageing. Apart from forestry, there are other activities, including a restaurant and spa business aimed at transit tourists on the national route. Nishiawakura had participated in discussions for mergers with the six neighbouring municipalities since 2000 but chose to be an independent village, as evident from the August 2004 resident questionnaire, which showed that 58.3 per cent were opposed to the merger (Seki, 2010).

The Centrum Woods Vision initiative

In 2007, the village established the Nishiawakura Employment Measures Council with the support of funds from the Ministry of Health, Labour and Welfare. This was to reconsider the village development and attract outsiders and youth. In response, the village office took the initiative to hire and develop human resources and set up incubation facilities to encourage new entrepreneurship. It was necessary to show the resolve to outsiders to bring potential human resources to the village. The 'Centrum Woods Vision' (2008) initiative was conceived jointly by the village office, the forestry cooperative and a collaborating consultant company. A declaration was taken to preserve and manage the trees planted in the village 50 years ago for the benefit of future generations. In addition to giving added value to the woods already planted, the village also aimed to create new woods that could fully underpin ecosystem services, with appropriate management to let daytime light reach the forest floor to foster forest bed vegetation and biodiversity. The woods in Nishiawakura are mostly privately owned, but as mentioned before, they were neglected due to the downturn in forestry and lack of proper management. To revive the dormant forestry sector, the village council decided to sign a 10-year business management agreement with wood owners and the forestry cooperative. This was to allow the implementation of the management plan that the village had formulated, and to perform thinning and build forest work roads for proper management. The profit from the thinning was to be shared between the village and the owners (Yokoe, 2017). It was in fact the first time that any municipality in Japan had taken over the management of privately owned woods on a long-term basis. Stakeholder engagement from the beginning helped to bring most of the 1,300 wood owners in the village to a consensus. The village office organized about 50 briefing sessions for each district to involve more consigners (Maki, 2018).

The village launched Japan's first-ever forest and forestry support project fund, 'Shared Forest' to attract initial investment for the initiative. An anonymous investment union purchased high-performance forestry machines, which were then loaned to the forestry cooperative. The union pays dividends mainly using rental expenses. The village succeeded in obtaining funding from 400 investors who were persuaded by the Centrum Woods Vision. During the project, village tours for investors were conducted to show the outcomes of their investments. As of March 2019, there were 778 consigners, which is a positive indication showing that the initiative is becoming popular.

Establishment of 'the School of Woods'

Previously, timber cut by the forestry cooperative had only been shipped to the market. To achieve the 'Centrum Woods Vision' objective, it was necessary to have an enterprise that added value to thinned woods from the forest management process. A pioneering and innovative development was triggered by a staff member of the forestry cooperative who started his own furniture workshop and started making furniture for infants and children using thinned woods from the village to visualize the vision. Inspired by this development, the 'Nishiawakura, the School of Woods' was established in 2009 with the aim of processing and distributing thinned woods and serving as a receptacle for people moving to Nishiawakura. Its main task was to purchase thinned woods from the forestry cooperative, lumbering and distribution.

To avoid competition with existing firms in the industry, the school started developing flooring products for urban residents and companies, that are sold directly online. It also lumbered small volumes of wood to meet the specific demands of small builders. The process did not go as smoothly as hoped, and the company faced several difficulties. However, consistent efforts to avoid subcontracting for large firms and of product differentiation led to its survival. Figure 9.1 shows how the thinned woods were distributed in Nishiawakura. The School of Woods purchased 5000 m^3 of thinned woods from the 'Centrum Woods Vision' project. Among them, class A and B materials were lumbered at the firm, and class

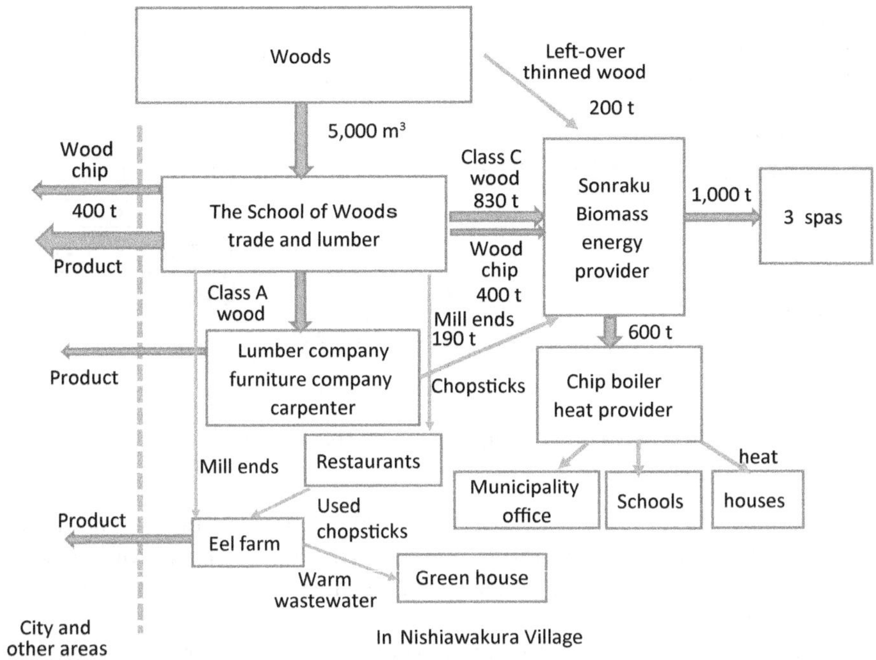

Figure 9.1 Value chain enhancement of woods in Nishiawakura.

C materials were sold to other firms in the village or to other areas. In-house lumbering and processing in the village spawned mill ends and chips which could then be used to create more added value in the village. The success of this firm was key to the development of new value chains. As of 2017, the firm has grown with 30 employees, of whom 14 are full-time workers. This shows how innovative ideas can help to improve the value chain and promote bioeconomy development in the rural areas. The direct benefits are the job creation and efficient use of wood resources and residues from the village forests which were otherwise unused.

Upstream development of a hundred-year forest business

The 'Centrum Woods Vision' initiative led to the start-up of new firms in woods management, woodwork and energy. Figure 9.2 shows the category of firms which were started in Nishiawakura in recent years. Three new woodworks have been launched in the village since 2007. Two of them produce furniture from cedar and cypress wood thinning, and the second one produces wooden toys. They all supported the vision and established workshops in the village. Nowadays, woodworkers in general in the region use materials of any origin but prefer to use Nishiawakura woods exclusively.

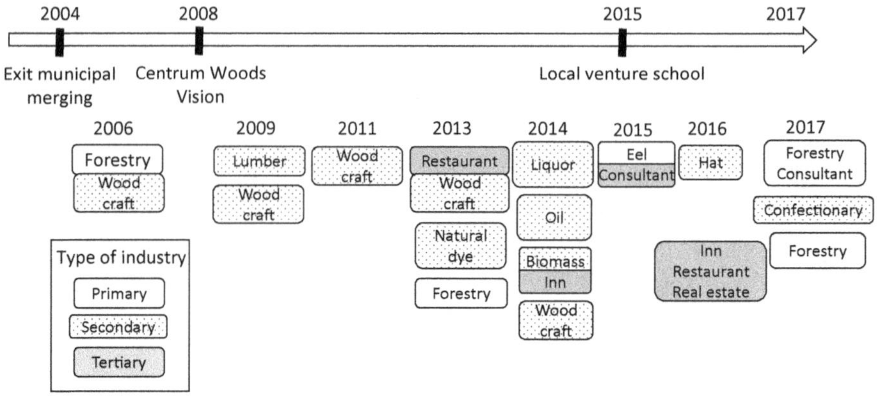

Figure 9.2 Chronicle of events/start-up firms in Nishiawakura.

Thus, a new market demand for the local wood has been generated. Although there is a certain brand power in the vision, each company is recognized in the market for their own product development and sales efforts.

In addition, three firms in logging and one firm in forestry management planning have started business in the region since 2013. The latter mainly has subcontracts with the 'Centrum Woods Vision' project. The founders of these firms are graduates of the entrepreneur school set up in the village, which will be described later in the chapter. As a result, important roles of the forest consignment management business of the municipality are now privatized. The motivation for the entrepreneurships resonates very well with the objectives of the vision. However, it is not easy for newcomers to gain the trust of wood owners. Involvement of the municipality in the business works as a kind of guarantee in this case.

Energy business enterprise

As part of the vision, the village of Nishiawakura is simultaneously aiming to pioneer in energy self-sufficiency. Joint solar power generation, small-scale hydropower generation and biomass heat generation projects are underway. Three small-scale hydroelectric power generators have already been installed in the village, which generate a total of approximately 300 kW. They use some of the local stream water, channelled through a pipe and generate power using the height difference of the steep slopes. The new unit is set to start operation in 2020, which will generate an additional 200 kW of energy. Currently, the generated electricity is being sold to the power company by Feed-in Tariff (FIT). Energy generated locally also meets 40 per cent of the village's electricity demand.

For biomass heat generation, class C thinned-wood materials are processed into 1 m-long wood fuel for water-heating at three hot spring facilities in the village (see Figure 9.1). A private start-up firm, namely, Sonraku Company, took charge of the operation as a heat provider in 2012. It also runs an inn equipped with one of the three hot springs. One thousand tonnes of firewood are used every year for the three facilities. In addition to this, in 2018, thermal energy supply to the administrative office facilities and nursery schools started, using an additional 600 tonnes of class C materials (see Figure 9.1). In all these cases, the village made the initial investment in the heat supply facility, and the Sonraku Company undertook heat supply operations. As of 2017, Sonraku has 19 employees, of whom 9 are full-time employees. This shows how innovative biomass use and innovative partnerships can generate new jobs in the rural areas, without causing negative impacts on the environment.

A start-up eel aquaculture project

The company A0, which specializes in consultancy business as a spin-off from the School of Woods, entered the eel-farming business in 2016. Grilled eel is a traditional Japanese delicacy with high market value supplied by many existing farms. An important component in farming eel is the maintenance of the water temperature in farms located in cold regions. The ideal water temperature for eel farming is 25–30 °C, and thus many eel farms are located in warmer areas in Japan. Nishiawakura village has a comparatively colder climate and thus requires water heating in the eel farms. A0 saw the potential to use the abundant local wood waste resources to warm the water in the eel tanks and thus established the eel production unit in Nishiawakura.

The waste wood materials generated by the School of Woods is used as fuel for the biomass boiler to heat 120 tonnes of water used for eel aquaculture. The closed-loop water recycling system implemented by A0 enabled minimizing water use as well as heat demand for the aquaculture since recycled water is much warmer than well water. The backwash water to clean the sand filters of the recycling system contains rich nitrogen from eel waste and thus is used to fertilize the soil in the nearby vegetable garden. The A0's eel farming unit is thus attempting to create a circular production system connecting forestry, aquaculture, and agriculture, which is a challenge. Experience so far indicates that it may take a few more production cycles before the process can be standardized.

Eel farming is dependent on capturing young eels from the wild (glass eels) caught in the estuary. The eel catch in general has been decreasing since 1982, and in 2014, the International Union for the Conservation of Nature (IUCN) designated the Japanese eel (*Anguilla Japtoa*) and the American eel (*Anguilla Rostrata*) endangered species 1B. In response to this, the Japanese government deployed permitting regulations and set an upper limit on the amount of eel aquaculture.

A0 is thus seeking a way to make eel farming sustainable and comply with the Aquaculture Stewardship Council (ASC) standards applicable for other fish species. Also, A0 is conducting field research to study the efficiency and impact of restocking of eels to the wild river thus helping in conservation of the eel.

Policy support for local initiatives

Nishiawakura has been supporting aspiring entrepreneurs with the 'Community revival cooperation corps' system, a scheme implemented in 2009, by the Ministry of Internal Affairs and Communications. The scheme renders people who transfer their residential registry to disadvantaged regions, eligible for support from the local government for a maximum of three years, through a grant of 2 million yen/year as a salary and 2 million yen/year for activities. Local governments can assign duties to the newcomers such as support of residents, sale and marketing of local brands and products, as well as working in agriculture, forestry and fisheries (Ministry of Internal Affairs and Communications, 2019). This scheme is effective in encouraging people to move to depopulated areas, lowering physical and psychological barriers by creating job opportunities and allowing time to adapt to the different lifestyles. Although this system is usually used to support local activities, entrepreneurs in Nishiawakura are allowed to spend all of their time and budget to develop their own businesses.

In 2015, a 'local venture school' was started as the village office's new project with the local consultancy company A0, the actual operator of the mission. The school's aim is to offer training to people inside and outside the village who want to be entrepreneurs. After screening, mentors advise entrants how to design their business through many presentations. The ones who reach the final selection are appointed as community revival cooperation corps and get support for three years if they move to live in the village. Two entrepreneurs were selected in 2015, four in 2016 and one in 2017 who moved and started business initiatives in the village. Though it is a small number, it shows that such incentives can motivate young people to move back to villages and help not only to revive the local economy but also add to the demographic change in a positive way. More such initiatives and incentives are necessary in the future to attract youth from urban areas to move to rural regions of Japan. As mentioned earlier, the necessary infrastructure for school education, health and energy requirements of the new residents need to be taken care of to keep their motivation.

Although the Centrum Woods Vision attracted wood resource-related entrepreneurs in the beginning, the types of new business have gradually diversified (see Figure 9.2). The local venture school furthered this phenomenon, bringing food oil production and sales, millinery, strawberries, a cake shop, restaurant business, plant dyeing, real estate business, to name but a few. Many entrepreneurs decided to live and start a business in the village because of the close support from the village office, which is important to motivate and support young entrepreneurs. The success of many predecessors was another aspect that attracted the new entrepreneurs (Bando, 2018).

Outcomes of the development

It is 15 years since Nishiawakura Village exited the municipality merger talk in 2004. The village has shown an extraordinary development since then. Achievements so far are multi-dimensional, and it can be better understood if summarized in relation to their contribution to several SDGs.

Change in population trends

The depopulation trend of Nishiawakura reversed in 2018. In 2017, 25 more people moved into than out of the village. The number of children under age 15 increased from 126 in 2011 to 155 in 2018 (Maki, 2018). The child-raising population in their twenties and thirties increased due to business start-ups and job postings that played a significant part. In Japan, entering nursery schools is often hard in urban areas due to lack of capacity, yet in rural areas it is relatively easy as there is enough room. It is also much easier for both parents to work while the children are at school.

In rural areas of Japan, it is not easy for new residents to rent facilities that are idle because of lack of trust. In Nishiawakura, the village office has been actively managing vacancies and idle properties, persuading absentee owners to rent them to newcomers and entrepreneurs. Because of the village's small size, it is easier for village officers to collect information or encourage property owners to rent out their properties. Proactive intervention by the village office helps to develop trust. At present, there are no idle facilities and new workshops and homes are being built, boosting demand for locally produced timber. Even a new private real estate business was launched recently.

Diversifying employment opportunities and work styles

In Nishiawakura, increasing job opportunities since 2007 has enhanced diverse work styles. It takes time and effort to process natural resources effectively. In the wood processing factory, the production line of lumber is deliberately separated into several stages. This enables housewives to work for a few flexible hours, a completely opposite strategy to mass production. Similarly, retired elderly people are working part-time for the heat provider. Production and transport of woods for boilers takes a lot of patience which is not always attractive to younger generations. In 2014, a retiree from the village office formed a non-profit organization, providing a new workplace for the disabled, which was new for the village. It undertakes activities such as product packaging, cloth-cutting as well as filter-cleaning for the eel culture. These are significant initiatives addressing Goal 5 (gender equality) and Goal 10 (reduced inequality) of the SDGs. Thus, the various initiatives consider engaging different age groups, building community trust and relationships, and at the same time managing the bioresources effectively.

Carbon absorption

Located in a mountainous area, Nishiawakura has been creating new value chains and increasing its use of renewable energy to use wood resources more effectively. Table 9.1 shows an estimation of the carbon emissions in Nishiawakura in 2011 and its projected goal in 2050. In 2011, 21,170 t CO_2/y was fixed in the village (emission: 13,135 t CO_2/y, fixation: 34,305 t CO_2/y). The village aims to fix 28,955 t CO_2/y by decreasing emissions by 2050 which is a 37 per cent improvement (Nishiawakura Village Council, 2013). Since the Centrum Woods Vision has set sector-specific numerical targets, long-term improvements not affected by personnel change is possible. Therefore, Goal 7 (affordable and clean energy), Goal 12 (responsible consumption and production), Goal 13 (climate action), and Goal 15 (life on land) of the SDGs are addressed simultaneously. Thus, Nishiawakura sets a good example of how the carbon emissions can be monitored at the sector level and set targets at the village level to cut down emissions and address SDGs. Lessons learnt from this case are relevant not only to Japan but also to other countries experiencing similar downtrends in rural development.

Table 9.1 Estimated carbon dioxide emission of Nishiawakura and future plan

		Unit: t-CO_2		
		2011	*2030*	*2050*
Fixation	Forest	–34305	–34305	–34305
	Households	3563	2400	1400
	Enterprises	2657	1000	900
Emission	Industry	1376	1300	850
	Transport	5539	3800	2200
	Subtotal	13135	8500	5350
Total		–21170	–25805	–28955

Source: Based on data collected from the village.

Attempt to link ecosystems and local ventures

An eel culture unit is normally located on the coast. Today's consumers are not aware that natural eels go up the river to mountainous areas and that a healthy river environment is essential for sustainable eel fishery. The attempt to raise eels in the mountains may not be efficient but is important. The idea of a mountainous village pursuing Goal 14 (life below water) is enlightening and innovative.

The bioeconomy effort is a pursuit of Goal 9 (industry, innovation and infrastructure). Nishiawakura, which aimed at regional revitalization based on its natural resources, is a good example of bioeconomy development. Early newcomers were indeed attracted by its vision. Gradually, the village is spawning more diverse entrepreneurs despite big risks when starting new businesses. What they require is reliable mentoring, trust and support from the local community. For this reason, entrepreneurs tend to gather in certain areas of the city where they can communicate with their peers. In the city, however, work and life are usually separated. Nishiawakura is attracting people because it can offer excellent social resources to entrepreneurs, both for living and working. The countryside is often attractive to live in with easy access to raw materials but may not be the best place for business. Facilitation measures like local venture schools are effective tools to attract new people.

Quick administrative decisions and project formations

Many municipalities in Japan operate focused only on tax revenues and grants from the state. Decision-making is usually very slow compared to the private sector. The Nishiawakura village office, on the other hand, procures various sources of funds through public offers, government initiatives and private subsidies. The village's strong determination to realize its vision and a close partnership with private consultants are the driving force. The larger the municipality, the more difficult it becomes to have a unique vision and differentiate it from others. Small municipalities, on the other hand, have a higher sense of responsibility shared by each officer, as well as room for decisions.

Nowadays national subsidies are shifting to competitive forms upon propositions, rather than uniform grants. Increasingly, private sector agencies are providing more funds to fulfil their corporate social responsibility, SDGs and other commitments. In that regard, municipalities need to be more strategic, transparent and quick in decision-making, establish links and partnerships as illustrated by Nishiawakura.

Conclusion

A series of development initiatives in Nishiawakura stemmed from the Centrum Woods Vision because it was a declaration of what ecosystem services the village was going to rely on 50 years ahead and until then. It visualized target material flow and appropriate size of business to be developed. This is an important lesson for planning bioeconomy at the village level, since potential biomass production is dependent primarily on the natural productivity of the region. In the fossil-oil-based economy, value chains were created based on economic efficiency. Mass production is effective at lowering the cost of products, thus mass producers attract suppliers and buyers around them to form an industrial ecology. Bioeconomy is far more dependent on regional ecosystem services and for its sustainability, the regional ecological footprint needs to be decreased and ecosystem services or biological productivity of the region need to be increased. Industrial ecology needs to be carefully planned to interact with and increase the ecosystem services of the region. This is a major difference from the current industry-based approach.

It is the municipality that took the initiative in Nishiawakura and brought investments to realize its vision. However, this may not be always the case for other municipalities. Addressing the SDGs also demands government and enterprises set their own goals and make improvements as a step towards sustainable development. *What are the ideal conditions and policy options which enhance such movements?* First, carbon footprint and ecosystem services need to be assessed at the regional or local scale. The former is relatively easier than the latter since some of ecosystem services are qualitative than quantitative. Quantification of these indices will clarify the status and goals to be achieved. If such assessment is carried out regularly throughout the nation, the design of subsidies or inter-regional trade based on commitments and actual achievements becomes possible. The Japanese government decided to impose a Forest Environment Tax from 2020, which implies that each taxpayer will be charged an additional 1,000 yen (approx. US$9) annually for better conservation and management of forests. The annual income would amount to 62 billion yen or US$563 million. The present design is aimed at consolidation of abandoned woods and management by the government to improve their ecosystem services. It is desirable to simultaneously allocate an adequate budget to establish assessment criteria for the above-mentioned indices.

Although ecosystem services are region-specific, the difficulty of development today is that it is hard to avoid competition. Nishiawakura's strategy was to differentiate itself from existing industries and avoid competition. This is what forced Nishiawakura to hold entrepreneur training sessions, improving the survival rate of newcomers to the village and developing the trust between partners before receiving them. Amenities for the residents increased because of the diversification of industries and the increase in the young population. Although its comprehensive approach can be replicated in other municipalities, the chance of success seems to remain stakeholder-dependent.

In 2018, the Cabinet Office started to give support to municipalities, which proposed distinctive strategies to pursue SDGs. Twenty-nine municipalities in 2018, and 31 (including Nishiawakura) in 2019 were chosen as model municipalities. Their accomplishments will be evaluated three years from now. Lessons from successful cases such as Nishiawakura, and failure from diverse cases with different ecosystem and societal background will provide good experiences for the ones who will follow them in future.

References

Bando, Y. (2018) 'Added value of new business in middle and mountainous areas', Master's thesis, Graduate School of Agricultural Science, Kobe University.

Cabinet Office (2014) 'Town, people, work creation law', (in Japanese). Available at: www.kantei.go.jp/jp/singi/sousei/info/pdf/konkyo1.pdf

Cabinet Office (2019) 'Bio-strategy 2019'. Available at: www.kantei.go.jp/jp/singi/tougou-innovation/pdf/biosenryaku2019.pdf

Duraiappah, A.N., Nakamura, K., Takeuchi, K., Watanabe, M. and Nishi, M. (2012) *Satoyama-Satoumi Ecosystems and Human Well-Being: Socio-Ecological Production Landscapes of Japan*. Tokyo: United Nations University Press.

FAO (Food and Agriculture Organization) (2018) 'FAO yearbook. Fishery and Aquaculture Statistics 2016'. Available at: www.fao.org/fishery/static/Yearbook/YB2016_USBcard/navigation/index_intro_e.htm

Fisheries Agency (2017) 'FY2016 trends in fisheries, FY2017 Fisheries Policy, White Paper on Fisheries: Summary'. Available at: www.jfa.maff.go.jp/e/annual-report/attach/pdf/index-7.pdf

Forestry Agency (2019) 'White Paper on forest and forestry 2018' (in Japanese. Available at: www.rinya.maff.go.jp/j/kikaku/hakusyo/30hakusyo/zenbun.html

Japan Policy Council (2014) 'Stopping the fall of birth rate: A strategy to revive the countryside'. Available at: www.policycouncil.jp/pdf/prop03/prop03_digest.pdf

Maki, D. (2018) 'Local venture', *Mokurakusha*.

Masuda, H. (2014) 'Depopulation and challenges to suburban regions'. Available at: www5.cao.go.jp/keizai-shimon/kaigi/special/future/0130/shiryou_09.pdf

Ministry of Economy, Industry and Commerce (MEIC) (2018) 'The 5th Strategic Energy Plan'. Available at: www.meti.go.jp/english/press/2018/pdf/0703_002c.pdf

Ministry of Environment (2017) 'Annual Report on the Environment, the Sound Material-Cycle Society and Biodiversity in Japan, 2016'. Available at: www.env.go.jp/policy/hakusyo/h28/html/hj1601030201.html

Ministry of Internal Affairs and Communications (MIAC) (2010) 'Grand municipality merger in Heisei'. Available at: www.gappei-archive.soumu.go.jp/heisei-nogappei.pdf

Ministry of Internal Affairs and Communications (MIAC) (2019) 'Community revival cooperation corps'. Available at: www.soumu.go.jp/main_content/000610488.pdf

Nakashizuka, T. and Kikuzawa, K. (2019) *Humans and Change of Forest*. Bun-ky: Kyoritsu Publishing.

Nishiawakura Village Council (2013) 'Action plan of Nishiawakura Village as an environmental model town'. Available at: www.vill.nishiawakura.okayama.jp/wp/wp-content/uploads/2014/03/20140404164032_2.pdf

Nishiawakura Village Council (2019a) 'Centrum Woods Vision'. Available at: www.vill.nishiawakura.okayama.jp/wp/%E7%99%BE%E5%B9%B4%E3%81%AE%E6%A3%AE%E6%9E%97%E6%A7%8B%E6%83%B3/

Nishiawakura Village Council (2019b) 'Statistics of Nishiawakura Village'. Available at: www.vill.nishiawakura.okayama.jp/wp/news/%E8%A5%BF%E7%B2%9F%E5%80%89%E6%9D%91%E3%81%AE%E7%B5%B1%E8%A8%88/

Seki, M. (2010) 'Nishiawakura Village/Centrum Woods Vision trial towards independence, Tiiki Kaihatsu (Area Development)', *Japan Center for Area Development Research*, vol. 545, pp. 56–58.

Yokoe, Y. (2017) 'Centrum Woods Vision of Nishiawakura Village', *Jititaihoumukennkyuu* [Municipality legal work research], vol. 48, pp. 27–29.

World Wide Fund for Nature, Japan (WWF Japan) (2010) 'Japan ecological footprint report 2009'. Available at: http://d2ouvy59p0dg6k.cloudfront.net/downloads/wwf_efj_2009e.pdf

10 Promoting sustainable biogas technology development through integrated food and energy systems

Roar Linjordet, Divina Gracia P. Rodriguez, Mehreteab Tesfai, Anne Strøm Prestvik, Salome Modiselle, Primrose Magama and Mokhele Moeletsi

Introduction

Energy is one of the most fundamental inputs to sustainable economic development. The UN Sustainable Development Goal 7 (SDG 7) therefore aims to ensure access to affordable, reliable, sustainable energy for all. It further emphasizes increasing the share of renewable energy in the global energy mix. Developing and supplying affordable and sustainable renewable energy services have continued to be a critical challenge to developing countries for various reasons (Mengistu et al., 2016). Achieving the targets set by SDG 7 by 2030 will be possible only if the current challenges are addressed immediately.

In Sub-Saharan Africa (SSA), 600 million people lack electricity and 890 million people cook with traditional fuels such as firewood, charcoal, cow dung and crop residues (IEA, 2018). Unsustainable felling of trees for firewood and for agricultural expansion is causing serious deforestation problems and land degradation (Rupf et al., 2015). Heavy reliance on traditional wood biomass as a source of fuel can be detrimental to the environment, and hinders social and economic development of rural communities in the long run. Access to clean and renewable energy will improve human health and well-being, and allow women and children to lead better lives (UN, 2018).

There is huge potential to provide clean and sustainable energy in SSA. A combination of wind, solar, biochar and biogas from agricultural residues will contribute significantly to energy supply in SSA in the future. Energy from biomass appears to hold the greatest potential for renewable energy (IEA, 2018). Bioenergy already provides half of all renewable energy that is consumed, and the IEA expects that modern bioenergy will continue to lead the field because of its huge potential. Because bioenergy sustainability is a complex issue, each application must be judged on its own specific circumstances, and generalizations regarding the sustainability of bioenergy feedstocks, fuels and technologies can be misleading. Biomass, which originates from plants and animals, will emit CO_2 when used for energy purposes. Sustainably managed fuel crops or waste biomass that otherwise are going to be burned or rot while giving off both methane and nitrous oxide will arguably not produce any net greenhouse gases (GHG) (Tenenbaum et al., 2014). A variety of bioenergy products, such as biochar, biogas, biofuel or biodiesel can be produced from renewable biomass. However, there is a lack of information regarding the framework conditions required to increase access to climate-smart energy while enhancing food security in a sustainable way for smallholder farmers. Moreover, SSA is witnessing increased frequency and intensity in climate change-associated extreme weather events, thereby increasing water, food, and energy insecurity. To meet these challenges, it is necessary that smallholder farming systems in African countries must undergo significant transformation (FAO, 2017). In this chapter, we will discuss how biogas technology can contribute to this transformation, which has the potential to enhance energy security. Biogas, suitable to smallholders, has become more attractive in the recent years because of the multiple benefits (environmental, economic and social) it provides and its potential contributions to several UN Sustainable Developmental Goals (Table 10.1).

Table 10.1 Multi-benefits of sustainable biogas development and contributions to relevant SDGs

Multiple benefits	*SDGs*	*References*
Environmental benefits		
• Promotes nutrient cycling of crop residues, manure	15	IRENA (2017)
• Reduces food spoilage	2	WBA (2013)
• Reduces release of greenhouse gases	13	Grieshop et al. (2011)
• Biogas stoves reduce use of wood or charcoal	15	Kammila et al. (2014)
• Prevents deforestation and improves land restoration	15	Subedi et al. (2014)
Economic benefits		
• Low cost technology both in capital and operation	7,8	Kammila et al. (2014);
• Increases resource use efficiency	12	Putti et al. (2015)
• Reduces wastes along the value chain	12	WBA (2013)
• Promotes circular economy/bio-based economy	8	IEA (2018)
• Potential for job creation in agriculture/forestry	8	IRENA (2017)
• Reduces need to buy chemical fertilizer	2	Bogdanski (2012)
Social and health benefits		
• Provides access to basic services	4	IRENA (2016, 2017)
• Improves health, reduces pneumonia and associated diseases	3	Smith et al. (2013, 2015)

Multiple benefits	*SDGs*	*References*
• Removes rotting organic waste and associated pathogens	6	IRENA (2017)
• Improves livelihoods of rural households	1	Bogdanski et al. (2010)
• Reduces time women spend collecting fuel wood and cow dung	5	Kammila et al. (2014)

The biogas production process, also termed anaerobic digestion (AD), take place in a closed chamber where all nutrients are preserved and transformed into a product called digestate that is rich in nutrients readily available for plant uptake, making the digestate a valuable organic fertilizer that is easy to store, transport and use (Box 10.1).

Biogas is unique compared to other renewable bioenergy sources in its versatility and flexibility in scale and design. It also benefits other priority areas such as sanitation, waste treatment and food security (Rupf et al., 2015). The biogas technology can be applied at various scales, from household domestic consumption to large-scale industrial applications. The biogas technology has been thoroughly studied and exploited in Europe and Asia since the 1980s as evident from the progress made in many countries. However, in Africa its adoption and dissemination is low and slow, despite the substantial efforts made by international organizations such as the Netherlands Development Organization (SNV) and the German Agency for Technical Cooperation (GTZ). The main contributing factors to the low adoption of biogas technology in Africa can be attributed to:

1 lack of adequate feedstock and water availability;
2 high initial/upfront cost of installation and lack of investments;

Box 10.1 Main features of biogas

Biogas is a modern form of bioenergy that can be produced through anaerobic digestion or fermentation of a variety of biomass sources. These include, but are not limited to, livestock manure, food waste, agricultural residues, energy crops, sewage and organic waste from landfills. Biogas is a versatile fuel that can be used for cooking, heating, lighting, power generation, and combined heat and power generation, as well as, when upgraded to boost its methane content, in transport applications. It also can provide dispatchable energy to power grids so that a higher share of electricity can be generated from wind and solar energy. Biogas typically consists of 50–75 per cent methane, which provides its energy content, and 25–50 per cent carbon dioxide, which potentially can be captured and stored.

Source: Adapted from IRENA (2017).

3 lack of skilled labour for installation and operation;
4 limited awareness of and training in biogas applications;
5 inadequate policy support and extension services;
6 slow behavioural and social acceptance.

In this chapter, different measures and options on how to solve and deal with the various constraints are discussed. Access to renewable and affordable energy will contribute to the implementation of sustainable intensification of feed and food production as outlined below. The chapter uses results from a case study in the Limpopo Province in South Africa that demonstrates that biogas from sustainable feedstocks is the technology of choice for resource-poor smallholder farmers. The next step is to consider the efforts that are needed to make further progress in promoting biogas technology. The opportunities and challenges of biogas technology using the Integrated Food and Energy Systems (IFES) approach are highlighted.

Literature review

In this section, a review and synthesis of recent research findings on the constraints on biogas production and adoption in Africa are presented, supported by examples from South Africa.

Feedstock availability for biogas production

Feedstock availability is one of the most important factors in the successful development of biogas. Different feedstock materials have different digestion properties, which affect the amount, composition and quality of the biogas and fertilizer produced. The digestion of the feedstock will depend on the amount of dry matter present in the material, and the proportion of the solid material that can be digested by the microorganisms (bacteria and archaea) and turned into biogas.

Another study from South Africa by Muvhiiwa et al. (2017) showed that there was a pervasive perception that biogas technology works only for those areas with access to large numbers of feedstocks containing enough animal and agricultural wastes. However, Brown (2006) suggested that one or two cows or between five and eight pigs could produce sufficient amount of feedstock for a single digester to provide biogas for a four-person household. This will be equivalent to 30 kg of cow manure or 15 kg of vegetable waste per household per day (Smith et al., 2015). When investigating the technical potential for household biogas in rural Africa, ter Heedge and Sonder (2007) used the number of domestic cattle per household as one of the criteria, namely, three or four night-stabled cows. The biogas technology is compatible with livestock farms since cattle are a kind of biogas reactor providing both microorganisms and feedstock for technical bioreactors.

Water availability for biogas production

Water is a key factor in the implementation of biogas in dry areas of SSA. The functioning of biogas digesters can be limited in areas with low water availability (Bansal et al., 2017) as water is already becoming a limited resource in many parts of Africa. A study conducted in Ethiopia showed that 60 per cent of 700 biogas digesters surveyed were out of operation due to lack of water. To run a biogas digester efficiently, the time taken to reach the water source should be not more than 30 minutes (Eshete et al., 2006). Moreover, the household should be able to collect at least 25 dm^3 water per person per day (Orskov et al., 2014). In addition to feedstock supply from cow manure, access to water allows a total of 18.5 million domestic biogas installations at the household level in Africa, which is a potential lost otherwise (ter Heedge and Sonder, 2007).

Biogas digester types

A biogas system has several elements including the feedstock collection and storage, the receiving and mixing area, the digester or reactor, the gas holder and the digester residue storage. A range of biogas digester types exists, and their geographical and contextual applicability should be taken into consideration to optimize productivity. In a recent review by Mutungwazi et al. (2018), the in-situ cast concrete digester was considered the most suitable design for installation in the South African context. Lower-cost, prefabricated biogas digester units constructed of fibre, plastic or lightweight bags can reduce installation costs. The use of low-cost pre-fabricated digesters is also anticipated to have a positive impact on the adoption of the technology as in the case of China (Box 10.2).

Policy and institutional support to biogas technology

Policies on renewable energy targets, feed-in-tariffs (FiTs), utility quota obligations, tendering, grid access, competitive bidding programmes, and fiscal incentives are already in place in some African countries. However, implementation of these policies has been poor and, in some cases, unattractive to investors (Kemausuor et al., 2018). According to Muvhiiwa et al. (2017) South-African decision-makers have poor knowledge and awareness of biogas technology. Hence, an adequate regulatory structure to support biogas technology is still lacking (SABIA, 2015).

Constraints on adoption of biogas technology in SSA

The key issue for biogas technology in Africa is to understand why scaling-up has not occurred despite demonstration by several programmes about

Box 10.2 Low-cost household biogas technology in China

Household biogas digesters have been implemented in China since the 1970s and especially promoted through the Programme for the Development and Promotion of Biogas Utilization in Rural China since 2001. A crude estimate says that household digesters served one quarter of China's rural population in 2010. There are three integrated livestock-biogas systems that are attuned to local climatic conditions. The 'Three-in-one' model combines a pig pen and a toilet with a digester and is used in southern China where the climate is moist tropical. The 'Four-in-one' model combines a pig pen, a toilet and a greenhouse with a digester and is common in northern China where the climate is colder. The 'Five-in-one' model also includes a solar-powered barn, a water-saving irrigation system and a water cellar, and is recommended for the dry regions in north-west China. All systems serve three purposes: (1) improve household hygiene and prevent environmental pollution from pig pens; (2) provide energy for cooking and lighting; and (3) provide an organic fertilizer through the digestate.

Average investment costs of household biogas digesters were $600 in 2010, but both national, regional and municipal governments subsidize the investment. The annual revenue for the household is low, only $25, but through the Clean Development Mechanism, a household may get up to $26 of income from emission reduction.

Source: Adapted from Bogdanski et al. (2010).

the viability and effectiveness of biogas plants. One factor that hinders the expansion of this technology is the lack of awareness and understanding of biogas as a form of energy (ibid.). In SSA, countries, such as Kenya, Tanzania and Ethiopia, have implemented national biogas programmes targeted at the rural adoption of biogas. In Kenya, the adoption of biogas technology has been slow, owing to inadequate funds, poor infrastructure and a general ignorance of the technology among the rural people who otherwise could derive the greatest benefit from it (Muvhiiwa et al., 2017). In Ethiopia, the low costs of the traditional fuels like firewood, the high costs of installation of biogas digesters and lack of knowledge of biogas installation, operation, and maintenance hinder the adoption of biogas (Surendra et al., 2014). In South Africa, there is the potential for further expansion of biogas technology (Tiepelt, 2015), but the adoption is a complex, bundled and incremental process, dependent upon household, national and regional accumulations of technological capabilities. As stated by Murphy (2001), and still valid: 'Biogas technology absorption, therefore, cannot occur without the proper social, cultural, political and economic institutions to support adoption, dissemination and appropriate contextual innovation.'

The food-energy nexus and the IFES approach

The IFES approach addresses the link between access to energy and food security (the food-energy nexus). Globally, almost 3 billion people lack access to adequate, reliable, clean and affordable energy sources for cooking and heating, and of this the largest proportion is in Africa (IEA et al., 2019). Without proper access to energy, it is difficult to achieve food security (Brew-Hammond, 2012). Widespread access to sustainable energy sources to produce, process, preserve, and consume food will enhance both food security and nutrition (Bogdanski, 2012). Sola et al. (2016) documented empirical evidence that shows the influence of energy access on food security (Table 10.2).

Renewable energy from biomass, such as biogas, could supply clean energy for cooking, create new jobs and technical skills, improve agricultural productivity and reduce GHG emissions. This would be a core factor for developing the bioeconomy. According to the FAO, the bioeconomy involves: (1) sustainable production and use of renewable biomass; (2) use of enabling and converging technologies; and (3) integration of different sectors, such as agriculture, health and industry (FAO, 2018b). Moreover, sustainable energy sources, such as biogas, and sustainable agricultural inputs, such as nutrient-rich digestate, minimize the environmental impacts on food production while improving nutrition and socio-economic development. To improve food security, one option is to implement sustainable intensification. Sustainable intensification is defined as producing more outputs with more efficient use of all inputs through various intensification processes, while reducing environmental damage, building resilience, natural capital and the flow of environmental services (Pretty et al., 2011). A safe integration of food and energy production may be one of the best ways to improve access to national food and energy security and simultaneously reduce poverty in a climate-smart way (FAO, 2011).

Table 10.2 Influences of energy access on food security

Influences	*References*
• Energy access influences dietary choices and cooking practices	Brouwer et al. (1996); Brouwer et al. (1997); Chikaire et al. (2011)
• Poor access to cooking fuel leads to reallocation of household resources (i.e., cash, time) from food production and preparation, to fuel procurement	Chikaire et al. (2011); Brouwer et al. (1997)
• Lack of access to energy leads to switches to inferior energy forms (e.g., charcoal to fuelwood, or fuelwood to dung), thereby reducing agricultural productivity	Dewees (1989); Brouwer et al. (1997)

The concept of Integrated Food and Energy Systems (IFES) is not new and the integration of food and energy production at both small and large scales has shown successful results, notably in Asian countries (Bogdanski et al., 2010). IFES is an integrated system that is designed to meet both food and energy needs through the application of sustainable intensification (FAO, 2011). There are two types of IFES:

- *Type 1 IFES:* Food and energy production are integrated either by producing biomass for food and energy on the same land using multi-cropping systems or mixing annual crops with perennial crops, i.e. agroforestry (Bogdanski et al., 2010). There are many examples of simple type 1 IFES that have proved to be able to provide food, feed and biomass for energy without compromising ecosystem services at the local scale (Bogdanski, 2012). The key to such systems is diversity, either by growing a mix of crops and/or the inclusion of woody perennials, to provide biomass for food, feed and energy (Bogdanski et al., 2010). One example from Malawi is the intercropping of staple foods (maize, sorghum, millets) with pigeon peas. The pigeon pea is a 'multi-purpose' nitrogen-fixing plant that both delivers a protein seed for human consumption, leaves and stems for animals' fodder and dry stems as fuel for cooking (Me-Nsope and Larkin, 2016).
- *Type 2 IFES:* The adoption of agro-industrial technologies such as anaerobic digestion, to maximize the utilization, recycling and economic use of all by-products from food production is a more technically advanced type of IFES (FAO, 2018a) (Figure 10.1). Biogas production using cow manure in a crop-livestock agricultural system is a simple example of type 2 IFES and there are many examples of such systems in South-East Asia (Bogdanski et al., 2010). Such systems use livestock waste (or even human waste), sometimes combined with crop residue, as feedstock. The biogas is mainly used for cooking, while the digestate is used as an organic fertilizer, sometimes released to ponds where microalgae and other aquatic plants become fish feed, or it is mixed with crop residues to speed up the composting process (ibid.). When the digestate is valued as fertilizer, it has a positive impact on the cost-benefit analysis of the system (Mwirigi et al., 2014). Digestate can directly lead to increased yields and reduced use of synthetic fertilizers when used as fertilizer, which in many cases is the most efficient way of using manure in crop production (de Groot and Bogdanski, 2013). However, as nutrients become more available, the risk of leaching also increases (Bonten et al., 2014), calling for use of sustainable practices when applying fertilizer (Smith et al., 2014).

One drawback is the added workload for the farmer, especially as the systems become more and more complex and the need for training increases. In some cases, the implementation of IFES may reduce the productivity of

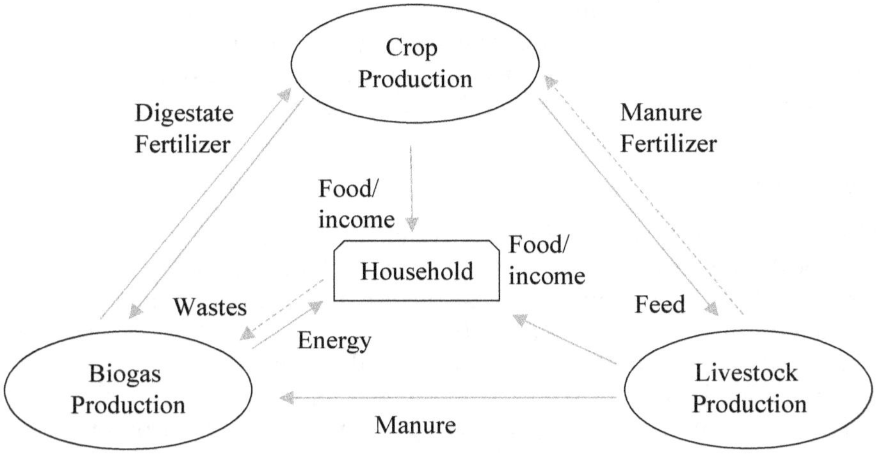

Figure 10.1 Schematic representation of type 2 IFES.

one of the components in the short term, e.g. lower yields of the main crop when interplanting with other crops (Bogdanski, 2012). The same applies to biogas systems where the investment cost is not covered by the energy replacement alone but may be viable together with higher yields from the use of the digestate as fertilizer (Kebedet et al., 2016). To assess the performance of the IFES system, all the elements of the systems, including the impact on the climate and the environment, as well as opportunities for trade, must be accounted for. For an IFES to be sustainable, all the components of the system need to work together. Moreover, the successful implementation of an IFES depends on a thorough understanding of the local context and the necessary capacity building among key stakeholders (FAO, 2018a).

The evidence of type 2 IFES working successfully in Africa is still very limited (Mwirigi et al., 2014). There is a lack of information in relation to the framework conditions required to increase access to climate-smart energy while enhancing food security in a sustainable way for small-holders in Africa. In the following section, the opportunities of biogas technology using IFES approach using the Limpopo Province in South Africa as a case study are presented. It must be emphasized that when the biogas intervention was conceived, it was more in the context of uplifting the rural household in a social context than for food security and profit maximization. The challenges to and opportunities in creating an enabling environment for the IFES approach are discussed and measures are suggested.

Case study background

South Africa is highly dependent on non-renewable sources of energy such as coal, oil and natural gas, making it one of the largest emitters of greenhouse gases in the world (ARC, 2015). On the other hand, renewable energy sources are currently contributing towards a small percentage of the country's total electricity demand. The country has abundant, diverse and unexploited energy resources (including biomass for biogas) that are yet to be used for improving the livelihoods of the population.

Site description

The study site is Malonga village in Makhado Local Municipality, one of four local municipalities falling under Vhembe District, one of the five districts making up Limpopo Province (Figure 10.2). The Makhado area is characterized by annual rainfall between 450–800 mm, however, high lying areas receive up to 2,300 mm of rainfall. Temperatures in the municipality range between 18 °C in the mountainous areas and 28 °C in the rest of the municipality. The average temperature is around 25 °C (MMIDP, 2014/2015). In Malonga village, some of the community members have boreholes in their yards and those who do not have these holes must fetch water from communal boreholes.

The aim was to implement a pilot biogas digester in Malonga rural community and evaluate biogas effectiveness and acceptance among the community. For this, a sample of 38 cattle-owning rural households was surveyed and their everyday livelihood strategies, including energy use,

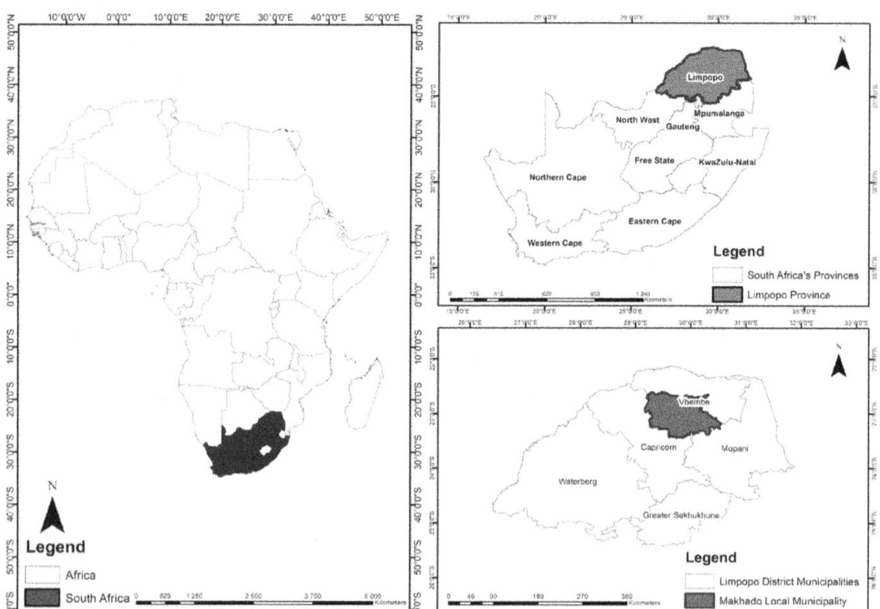

Figure 10.2 Maps showing case study location, globally, nationally and locally.

were monitored. The community demographics revealed that from the total sample of 38 household heads interviewed, 66 per cent were females and 34.2 per cent were males. Some 90 per cent of the household heads were unemployed and 10 per cent were employed either on full-time or part-time basis. About 92 per cent received monthly income from the social grants through the South African welfare system.

All the households indicated that they had access to land, 58 per cent had less than 2 ha of arable land while 42 per cent had 2 ha or more. All the households cultivated maize crop, and 24 per cent of the households practise intercropping maize with groundnuts. These households also produced vegetables such as Chinese spinach, beans, pumpkin, butternut, and lettuce. Twenty-nine households indicated that they used fertilizers such as nitrogen, phosphorus and potassium mix (NPK), lime, organic fertilizer and ordinary *kraal* manure. The reasons for not using fertilizers were due to high cost, long distances to markets and inaccessible to buy when needed. Seventy-nine per cent indicated that they did not hire farmworkers but carried out farming activities on their own. All respondents had *kraals* around their households, which are enclosures, in which the cattle are kept overnight.

As part of the study, installation of biogas digesters in three households was performed. Selection of the three households was based on number of cows owned (minimum 10), water availability (water source within the yard), and household energy demand. Three biogas digesters with 6 m^3 capacity each were installed in the three households. An AGAMA 6 m^3 prefabricated digester made of linear low-density polyethylene (LLDPE) was selected due to ease of installation, low skill levels required to install, low installation time (7–14 days) and limited materials required for installation. These digesters were filled with water and cattle dung from kraals around the households' yards.

Results of the case study

The case study results in relation to household energy consumption, firewood and cow dung collection, capacity building, digester installation cost, and economic benefit of biogas technology, are presented below.

Energy consumption

All the households (HH) sampled for the survey had electricity installed. The three main energy sources used around the house were electricity, firewood, and candles. Electricity is the main source of energy used for lighting, refrigeration, cell phone charging, TV, radio, cooking and water heating. Some 97 per cent indicated that their main source of energy for cooking was firewood, and the same number indicated that they used electricity for lighting. About 82 per cent of the households access electricity from Giyani town (20 km away from the village), and firewood and candle were sourced locally. Almost 55 per cent of the sample households indicated that the amount they spent on electricity was R150–R200[1] per month while the remaining 45 per cent spent R50–R120 on electricity in the same period.

Firewood and cow dung collection

It was found that more firewood is collected in winter than in summer and an average 5 hours were spent each time to collect firewood from the nearby forest that is between 3 and 5 km away from the households. The labour for firewood collection from the forest was done by a family member or hired person. The majority of the households collect dung using family labour. They used the dung to decorate the floors and walls of their huts. They collect the dung once from the kraal in the morning, which is about one bucketful, collected in half an hour.

Capacity building and training

In the village, the key stakeholders include the chiefs, and a marginalized group including youth and women, who are likely to contribute to the success of the project. The chiefs provide governance and help the marginalized groups of the community. The latter were trained in biogas technology by the study team, and it is expected this will improve their socio-economic status by opening up new job opportunities and financial savings. Stakeholder engagement and training workshops were conducted in the community to foster buy-in and participation of the community to install the biogas digesters. The interactive training workshops discussed issues related to climate change, introduced biogas digesters and demonstrated the operation and maintenance of biogas digesters. Skills transfer was an important outcome of the project and this was achieved through training local youth on how to install the biogas digesters. The youth took the responsibility to install the biogas digesters at the demonstration sites under the supervision of the project team.

Economic benefits from biogas technology

Anaerobic digestion products are biogas and digestate (Table 10.3). The household has to feed the digester 40 kg of fresh cattle manure and two buckets (20–24 litres) of water daily as specified by the manufacturer of the biogas digester (Agama, 2009) to produce 2 m^3 of biogas or 720 m^3/annum of biogas, about 310 kg of biogas equivalent LPG generation per annum. This was based on the assumption that 1^3 of biogas is equivalent to 0.43 kg of liquid petroleum gas LPG (Singh et al., 2014). A kg of LPG costs R24.44, hence 1 m^3 of biogas is priced at R10.51. The digestate

Table 10.3 List of parameters used to estimate biogas feasibility

	Unit price	*Daily*	*Annual*
Output			
• Biogas (m^3)		2	720
• Digestate (kg)	R0.05	80	28,800
• Biogas equivalent LPG (kg)	R10.51	0.86	310
Input			
• Fresh cow dung (kg)		40	14,400
• Water (litre)		40	14,400
• Labour (hour)	R15.00	1	360
• Maintenance and support (R)	R220.00		2,640

produced is 80 kg/day or 28,800 kg/annum. It is assumed that the average life-span of the biogas digester is 20 years with a discount factor (DF) of 10 per cent. Input costs include labour and maintenance and support services. Water is free, as all households have access to water. Maintenance and support were provided by a trained biogas technician from the community whose daily rate is R220/day. Technical assistance was provided to a household once a month. The biogas digester installation cost was approximately R43,000.

The feasibility analysis of biogas installation was based on the cash flow forecast (Table 10.4). A sensitivity analysis was done with two more scenarios to compare costs and benefits. The two new scenarios were based on the two-buckets scenario which has been approved by Agama (2009). In one scenario it is planned that maintenance and support costs will be stopped in the 11th year of the 20-year evaluation of the digester investment. It is expected that beyond the tenth year the household will be confident enough to take up the responsibility of maintenance without assistance. The second scenario is of 50 per cent increase in products prices. It is

Table 10.4 Analysis of criteria of biogas installation feasibility

Criteria of business feasibility	*Base scenario*	*50% increase in output prices*	*Reduction of maintenance costs*
Net present value (NPV) in Rand	19,484	101,430	33,589
Benefit-cost ratio (BCR)	1.1	1.7	1.3
Internal rate of return (IRR) (%)	21	36	21
Payback period (PP) (years)	5.25	3.5	5.25

envisaged that as renewable energy developments in the country increase, the renewable energy demand will also increase, hence 50 per cent increase of products prices.

The results showed that investment in biogas digester was feasible. The net present value (NPV) was greater than zero. This suggests that the present value of incremental benefit is greater than the present value of all investment and operating costs. The benefit-cost ratio (BCR) was also greater than 1, implying that for each R1.0 invested at a discount rate of 10 per cent, a return of R1.1 was obtained. The internal rate of returns (IRR) was 21 per cent, implying that the profit level of the net investment in biogas digester is larger than the DF. However, the household will only be able to pay back the digester capital outlay in 5.25 years of its operations. Results of the sensitivity analysis also suggested positive NPV > 0, BCR > 0 and IRR > DF. Nevertheless, the payback period does not change much, whether the output prices increase, or the maintenance costs are reduced. The biogas project is not feasible, however, if the household decides to produce only 1 m^3 of biogas daily, with NPV = -R64,887, BCR = 0.57, IRR = 9 per cent.

Results thus far support the findings from other developing countries that found biogas digesters can be a viable technology for rural households that have access to animal manure and a reliable water supply. Combustible biogas was produced within two months of installation and the three households included in the study were able to meet their cooking energy needs. The monetary savings from switching to biogas could not easily be quantified as the households were replacing firewood that they were collecting. Other benefits gained included *convenience of using biogas stoves, less time was needed to collect cow dung and, unlike firewood, cow dung collection was not affected by the rainy season.* The major challenges faced by the households were blockages of the inlet feed ports and outlet pipes which would cause the digester not to produce gas. With time, however, some of the households learnt how to unblock the port and outlet pipes. Another issue was limited use of the digestate as fertilizer because there were limited crop production activities due to limited water supply. Consequently, the IFES approach was not fully implemented in the study area, partly because the digesters were designed primarily to meet the households' cooking needs, and partly due to lack of water.

Discussion

Robust sustainability frameworks are key to bioenergy growth. Only bioenergy that reduces lifecycle GHG emissions while avoiding unacceptable social, environmental and economic impacts has a future role in decarbonizing the energy system. Robust sustainability governance and enforcement must therefore be a central pillar of any bioenergy support policy (IEA, 2018). While the IFES provides many advantages compared to conventional farming systems, it will only succeed in delivering its co-benefits (i.e. access to food and energy) to smallholder farmers, if the following framework conditions are in place:

1. *Access and availability of inputs (feedstocks, water)* – In the case study site, the labour spent on producing biogas can limit its use as a source of energy and fertilizers. Transporting water accounts for the largest share of labour input. Installation of a biogas digester could be coupled with rainwater harvesting technologies or borehole drilling to achieve successful implementation of IFES in the case study area. As it is not possible to subsidise individuals and organizations in South Africa who are using municipal water to feed digesters, a viable option is to install the biogas digesters at the sites where there is free water access. There are various low-cost water harvesting structures that can be used to supply additional water for domestic biogas productions (Mekdaschi and Liniger, 2013). These include domestic water recycling, rainwater harvesting (roof top/courtyard, open ponds, ground well, micro-catchment) and aquaculture. Whenever possible, gravity-fed closed pipeline systems can improve water availability and reduce farm and management costs. Technologies for greywater harvesting must be adopted to address water challenges/scarcity, increase biogas production, and save water for irrigation and food production. For example, rainwater harvesting tanks are already implemented with domestic biogas systems in some regions, such as Fada N'Gourma in Burkina Faso (Huba et al., 2007). In South Africa, greywater generated per household is approximately 75 per cent of the household water consumption (Carden et al., 2007).
2. *Slurry transportation is also another challenge* – Hence, the installation and operation of biogas digester offer a realistic option if the time spent on biogas production does not exceed the time spent for wood collection. Biogas can help save labour time when resources are available. For smallholders without access to trucks and tanks, one option is to convert slurry into compost with other organic wastes before carrying it to the fields to reduce the labour time and to recycle the nutrients. Applying different feedstock improvement solutions (e.g. nutrient addition, co-digestion and biomass pre-treatment) to feedstock has also been demonstrated to be an effective means of enhancing the methane yield of the feedstock, thereby improving the overall anaerobic digestion process (Nges, 2012).
3. *Access to knowledge and information* – The baseline study indicated that 92 per cent of the sample farmers did not adopt biogas digester simply because they were not aware of or not heard about the biogas technology. Farmers need to know about: (a) the availability of the technology; (b) its costs and benefits; and (c) the advantages/disadvantages of using them to be able to decide whether to adopt or decline a technology. The

awareness campaigns could include: (a) information on the available renewable energy sources and renewable power-generating technologies, including their economic viability, designs, construction, operation, and maintenance; (b) the potential benefits of IFES among rural communities; and (c) information on renewable energy policy commitments.

4. *Access to financing mechanisms* – IFES requires some form of investment, from investing in capacity building (e.g. use of biogas technology) to more substantial investment in technologies (e.g. installation of biogas digesters). The high upfront capital costs associated with biogas technology restrict uptake and adoption of the technology (Mulinda et al., 2013; Roopnarain and Adeleke, 2017). As in other African countries, such as Ethiopia and Uganda, the construction cost of a 6 m^3 biogas digester in South Africa is quite high (R43,000 or US$2,838). To motivate farmers and other private sector players to adopt biogas technology, monetary incentives through grants or subsidies to bring down financial capital cost must be considered (Rajendran et al., 2012; Giwa et al., 2017). Based on a feasibility study, the prospect of adopting biogas digesters at the farmer level is likely to be high. However, co-funding in terms of in-kind contribution should be introduced to sustain the technology implementation. Sustained provision of subsidies (e.g. financial assistance) cannot be paid on an on-going basis in the future. Linking financial institutions and developing financial products that will allow farmers improved access to loans from formal credit/financial institutions could replace subsidies and encourage the uptake of biogas technology. There is also a need to develop cheaper biogas digesters to expand the use of biogas digesters for households, e.g. water gas heaters, refrigeration and heating.
5. *Training and capacity-building* – Even if a government offers financial support and low-cost capital, farmers will not adopt IFES-based biogas if they lack the required complementary knowledge and skills. In the case study area, training and building technical capacity on biogas are needed to support farmers to enable wider adoption and dissemination of biogas technology. For example, the annual maintenance and support cost is estimated at about 6.14 per cent of the total construction cost of the biogas digester. Capacity-building can be done by setting up extension and service systems that provide training and help maintain the biogas systems already installed in the households. The biogas infrastructure companies can also provide training to farmers on quality assurance measures to enhance quality and accountability. This strategy should also include a set of accompanying measures to support households making productive use of biogas.
6. *Policy support and stakeholder collaboration* – In order for the development of a biogas technology to be successful, it is not only important to optimize the technology itself, but also to simultaneously create a

policy and institutional environment in which the technology has a place (Schot and Geels, 2008). This is done, for instance, by involving key stakeholders (from farmers to policy-makers) right from the conceptualization of the biogas project to the implementation phase in order to build up trust and ensure sustainability of the project outputs. Collaboration of farmers to operate a community biogas system may be an alternative option to provide clean and sustainable energy with acceptable labour requirements of production and reduce the implementation risks.

7. *Access to insurance and technical support services* – Farmers' risk aversion to biogas systems can be addressed through product warranties or risk insurance. For example, maintenance and repair costs of biogas digesters in the case study area are very expensive. Hence, repair not only affects current users, it also can dissuade potential adopters. Big commercial farmers and businesses have a variety of insurance instruments while smallholder farmers are mostly uninsured. They are exposed to risks and they rely on the limited technical training they acquired prior to the installation of the digester. Backstopping technical support and maintenance services are important for the sustainability of the biogas systems (Bogdanski et al., 2010). Smallholder farmers need access to reliable technical support services. The availability of technicians who can monitor, repair and maintain the biogas systems would increase the adoption rates and guarantee long-term utilization of biogas. The Kenya Biogas Program, for instance, required Biogas Construction Enterprises and biogas companies to provide (a) at least a one-year warranty for all systems installed, and (b) customer support/customer support centres to measure client satisfaction and plant functionality.

Conclusion

This chapter has reviewed and discussed the main challenges and opportunities to promote sustainable biogas technology development through integrated food and energy systems (IFES) in SSA in general and in particular in South Africa. The case study results from Malonga village in Limpopo Province were analysed with respect to technical and economic feasibility. In general, the investment in biogas digester installed in the case study was profitable in terms of NPV, BCR and IRR but care should be taken when applying the results to other areas. This is because our analysis has some limitations: (1) the study did not consider the cultural, social and environmental risks of adoption of biogas: and (2) the role of gender in biogas technology adoption was not assessed separately.

Nonetheless, based on the baseline information collected from the case study, an integrated framework condition suitable to the smallholder biogas technology was suggested. IFES will only succeed, if the following framework conditions are in place:

1. Access and availability of feedstocks and water.
2. Access to knowledge and information.
3. Access to financing mechanisms.
4. Training and capacity-building.
5. Policy support and stakeholder collaboration.
6. Access to insurance and technical support services.

The implementation of the framework conditions for biogas technology should be underpinned by conducting an integrated research using IFES type 2 approach in the context of the smallholder farmers in SSA.

Note

1 Currently, 1 US dollar is equivalent to 14.7 ZAR (South Africa Rand currency).

References

AGAMA Biogas (Pty) Ltd (2009) 'Sustainable cities: biogas energy from waste'. Cape Town: AGAMA Biogas (Pty) Ltd.

ARC (2015) 'Sustainable biogas production for South African rural households: Limpopo situational analysis'. Unpublished report.

Bansal, V., Tumwesige, V. and Smith, J.U. (2017) 'Water for small-scale biogas digesters in sub-Saharan Africa', *GCB Bioenergy*, vol. 9, pp. 339–357.

Bonten, L.T.C., Zwart, K.B., Rietra, R.P.J.J., Postma, R. and Haas, M.J.G. (2014) 'Bio-slurry as fertilizer: Is bio-slurry from household digesters a better fertilizer than manure? A literature review'. Wageningen: Alterra.

Bogdanski, A. (2012) 'Integrated food–energy systems for climate-smart agriculture', *Agriculture & Food Security*, vol. 1, no. 1.

Bogdanski, A., Dubois, O., Jamieson, C. and Krell, R. (2010) *Making Integrated Food-Energy Systems Work for People and Climate*. Rome: FAO.

Bogdanski, A., Olivier D. and Chuluunbaatar, D. (2010) 'Integrated food-energy systems: Project assessment in China and Vietnam, Final Report'. Rome: FAO.

Brew-Hammond, A. (2012) 'Energy: The missing millennium development goal'. In F. Toth (ed.) *Energy for Development*. Dordrecht: Springer, the Netherlands.

Brouwer, I. D., den Hartog, A.P., Kamwendo, M.O. and Heldens, M.W. (1996) 'Wood quality and wood preferences in relation to food preparation and diet composition in Central Malawi', *Ecology of Food and Nutrition*, vol. 35, no. 1, pp. 1–13.

Brouwer, I.D., Hoorweg, J.C. and Van Liere, M.J. (1997) 'When households run out of fuel: Responses of rural households to decreasing fuelwood availability, Ntcheu District, Malawi', *World Development*, vol. 25, no. 2, pp. 255–266.

Brown V.J. (2006) 'Biogas: A bright idea for Africa', *Environmental Health Perspectives*, vol. 114, pp. 301–303.

Cai, X., Magidi, J., Nhamo, L. and Koppen, B. (2017) 'Mapping irrigated areas in the Limpopo Province, South Africa', IWMI Working Paper No. 172. Available at: www.iwmi.cgiar.org/publications/iwmi-working-papers/iwmi-working-paper-172/

Carden, K., Armitage, N., Winter, K., Sichone, O. and Rivett, U. (2007) 'Understanding the use and disposal of greywater in the non-sewered areas in South Africa'. Water Research Commission Report, no. 1524/1/07. Cape Town: Water Research Commission.

Chikaire, J., Nwakwasi, R.N., Osuagwu, C.O. and Oparaojiaku, J. (2011) 'Renewable energy resources for women's empowerment in Nigeria', *Researcher*, vol. 3, no. 10, pp. 348–357.

de Groot, Lennart, and Bogdanski., A. (2013). 'Bioslurry = Brown Gold? A review of scientific literature on the co-product of biogas production'. Available at: www.fao.org/3/a-i3441e.pdf

Dewees, P.A. (1989) 'The wood fuel crisis reconsidered: Observations on the dynamics of abundance and scarcity', *World Development*, vol. 17, no. 8, pp. 1159–1172.

Eshete G., Sonder K. and ter Heegde, F. (2006) 'Report on the feasibility study of a national programme for domestic biogas in Ethiopia'. Available at: www.bibalex.org/Search4Dev/files/338849/172350.pdf

FAO (Food and Agriculture Organization) (2011) *Building Resilience through Safe Access to Fuel and Energy (SAFE)*. Rome; FAO.

FAO (Food and Agriculture Organization) (2017) 'Climate Smart Agriculture Sourcebook'. Available at: www.fao.org/climate-smart-agriculture-sourcebook

FAO (Food and Agriculture Organization) (2018a) 'Lessons learned on the sustainability and replicability of integrated food-energy systems in Ghana and Mozambique'. Available at: www.fao.org/3/i8627en/I8627EN.pdf

FAO (Food and Agriculture Organization) (2018b). 'Assessing the contribution of bioeconomy to countries' economy: A brief review of national frameworks'. Available at: www.fao.org/3/I9580EN/i9580en.pdf

Giwa, A., Alabi, A., Yusuf, A., and Olukan, T. (2017) 'A comprehensive review on biomass and solar energy for sustainable energy generation in Nigeria', *Renewable and Sustainable. Energy Reviews*, vol. 69, pp. 620–641.

Grieshop, A., Marschall, J. and Kandlikar, M. (2011) 'Health and climate benefits of cookstove replacement options', *Energy Policy*, vol. 39, no. 12, pp. 7530–7542.

Huba, E., Fall, P.A., Sanogo, O., Kabore, G. and Bracken, P. (2007) 'Feasibility study for a national domestic biogas programme in Burkina Faso'. Available at: https://energypedia.info/images/3/3e/Domestic_Biogas_Program_in_Burkina_Faso.pdf

IEA (2018) 'Renewables 2018 Analysis and Forecasts to 2023'. Available at: www.iea.org/renewables2018/.

IEA, IRENA, UNSD, WB, WHO (2019) 'Tracking SDG7: The energy progress report 2019'. Available at: https://sustainabledevelopment.un.org/content/documents/2019_Tracking_SDG7_Report.pdf

IFES (2019) 'Biogas: A renewable biofuel'. Available at: http://biogas.ifas.ufl.edu/feedstocks.asp

IRENA (2016) 'Measuring small-scale biogas capacity and production', Abu Dhabi: IRENA. Available at: www.irena.org/DocumentDownloads/Publications/IRENA_Statistics_Measuring_small-scale_biogas_2016.pdf

IRENA (2017) 'Biogas for domestic cooking: Technology brief'. Available at: www.irena.org/publications/2017/Dec/Biogas-for-domestic-cooking-Technology-brief

Kammila, S., Kappen, J.F., Rysankova, D., Hyseni, B. and Putti, V.R. (2014) 'Clean and improved cooking in Sub-Saharan Africa: A landscape report'.

Available at: http://documents.worldbank.org/curated/en/164241468178757464/Clean-and-improved-cooking-in-Sub-Saharan-Africa-a-landscape-report

Kebede, E., Gan, J. and Kagochi, J. (2016) 'Agriculture based energy for rural household income and well-being: East African experience', *Renewable and Sustainable Energy Reviews*, vol. 53, pp. 1650–1655.

Kemausuor, F., Adaramola, M.S. and Morken, J. (2018) 'A review of commercial biogas systems and lessons for Africa', *Energies*, vol. 11, pp. 29–84.

Mekdaschi, S.R. and Liniger, H. (2013) 'Water harvesting: Guidelines to good practice' Available at: https://wocatpedia.net/wiki/File:Water_Harvesting_Guidelines_to_Good_Practice.pdf

Mengistu, M.G., Simane, B., Eshete, G. and Workneh, T.S. (2016) 'Factors affecting households' decisions in biogas technology adoption: The case of Ofla and Mecha Districts, northern Ethiopia', *Renewable Energy*, vol. 93, pp. 215–227.

Me-Nsope, N. and Larkins, M. (2016) 'Empowering women for food and income security: The case of pigeon pea in Malawi'. Working Paper No. 308, Michigan State University, East Lansing.

MMIDP (2014/2015) 'Makhado Municipality Integrated Development Plan'. Available at: http://mfma.treasury.gov.za/Documents/01.%20Integrated%20Development%20Plans/201415/02%20Local%20Municipalities/LIM344%20Makhado/LIM334%20Makhado%20Final%20IDP%20Review%202014%20-15.pdf

Mulinda, C., Hu, Q. and Pan, K. (2013) 'Dissemination and problems of African biogas technology', *Energy and Power Engineering*, vol. 5, pp. 506–512.

Murphy, J.T. (2001) 'Making the energy transition in rural East Africa: Leapfrogging an alternative?', *Technological Forecasting and Social Change*, vol. 68, pp. 173–193.

Mutungwazi, A., Mukumba, P. and Makaka, G. (2018) 'Biogas digester types installed in South Africa: A review', *Renewable and Sustainable Energy Reviews*, vol. 81, pp. 172–180.

Muvhiiwa, R., Hildebrandt, D., Chimwani, N., Ngubevana, L. and Matambo, T. (2017) 'The impact and challenges of sustainable biogas implementation: Moving towards a bio-based economy', *Energy, Sustainability and Society*, vol. 7, no. 1.

Mwirigi, J., Balana, B.B., Mugisha, J., Walekhwa, P., Melamu, R., Nakami, S. and Makenzi, P. (2014) 'Socio-economic hurdles to widespread adoption of small-scale biogas digesters in sub-Saharan Africa: A review', *Biomass and Bioenergy*, vol. 70, pp. 17–25.

Nges, I.A. (2012) 'Anaerobic digestion of crop and waste biomass: Impact of feedstock characteristics on process performance', PhD thesis, Lund University, Lund, Sweden.

Orskov, B., Yongabi, K., Subedi, M. and Smith, J. (2014) 'Overview of holistic application of biogas for small scale farmers in sub-Saharan Africa', *Biomass and Bioenergy*, vol. 70, pp. 4–16.

Pretty, J.N., Toulmin, C. and Williams, S. (2011) 'Sustainable intensification in African agriculture', *International Journal of Agricultural Sustainability*, vol. 9, no. 1, pp. 5–24.

Putti, V.R., Tsan, M., Mehta, S. and Kammila, S. (2015) 'The state of the global clean and improved cooking sector', Technical Paper no. 007/15, ESMAP (Energy Sector Management Assistance Program). Washington, DC: World Bank. Available at: https://openknowledge.worldbank.org/handle/10986/21878 License: CC BY 3.0 IGO

Rajendran, K., Aslanzadeh, S., Taherzadeh, M.J., Rajendran, K., Aslanzadeh, S. and Taherzadeh, M.J. (2012) 'Household biogas digesters: A review', *Energies*, vol. 5, pp. 2911–2942.

Roopnarain, A. and Adeleke, R. (2017) 'Current status, hurdles and future prospects of biogas digestion technology in Africa', *Renewable and Sustainable Energy Reviews.*, vol. 67, pp. 162–179.

Ruffini, A. (2013) 'SA not using its biogas potential', *Africa Powers Journal*. Available at: www.esi-africa.com/top-stories/sa-not-using-its-biogas-potential/

Rupf, G.V., Bari, P.A., de Boer, K. and McHenry, M.P. (2015) 'Barriers and opportunities of biogas dissemination in Sub-Saharan Africa and lessons learned from Rwanda, Tanzania, China, India, and Nepal', *Renewable and Sustainable Energy Reviews.*, vol. 52, pp. 468–476.

SABIA (South African Biogas Industry Association) (2015) 'National Biogas Conference', Standards and Regulations, 5 March.

Schot, J.W. and Geels, F.W. (2008) 'Strategic niche management and sustainable innovation journeys: Theory, findings, research agenda, and policy', *Technology Analysis and Strategic Management*, vol. 20, no. 5, pp. 537–554.

Singh, R., Behera, S., Yadav, Y.K. and Kumar, S. (2014) 'Potential of wheat straw for biogas production using thermophiles', in S. Kumar, A.K. Sarma, S.K. Tyagi and Y.K. Yadav (eds) *Recent Advances in Bio-Energy Research*. Kapurthala: SSS-National Institute of Renewable Energy.

Smith, J., Abegaz, A., Matthews, R.B., Subedi, M., Orskov, E.R., Tumwesige, V. and Smith, P. (2014) 'What is the potential for biogas digesters to improve soil fertility and crop production in Sub-Saharan Africa?', *Biomass and Bioenergy*, vol. 70, pp. 58–72.

Smith, J., Apsley, A., Avery, L., Baggs, E., Balana, B., Bechtel, K., Davidson, G., et al. (2013) 'The potential of small-scale biogas digesters to alleviate poverty and improve long term sustainability of ecosystem services in Sub-Saharan Africa', Final Report, Department for International Development, Project DFID NET-RC A06502. Available at : https://assets.publishing.service.gov.uk/media/57a08a1740f0b652dd000566/60928-FinalReport140613.pdf

Smith, J., Austin, G., Avery, L. and Balana, B. (2011) 'The potential of small-scale biogas digesters to alleviate poverty and improve long term sustainability of ecosystem services in Sub-Saharan Africa'. Aberdeen: Institute of Biological and Environmental Science, University of Aberdeen.

Smith, K., Pillarisetti, A., Hill, L.D., Charron. D., Delapena, S., Garland, C. and Pennise D (2015) 'Proposed methodology: Quantification of a saleable health product (aDALYs) from household cooking interventions', Berkeley Air Monitoring Group. Available at: http://ehsdiv.sph.berkeley.edu/krsmith/%20publications/2015/aDALY_Methodology.pdf

Smith, K., Uma, R., Kishore, V.V.N., Zhang, J., Joshi, V. and Khalil, M.A.K. (2000) 'Greenhouse implications of household stoves: An analysis for India', *Annual Review of Energy and the Environment*, vol. 25, pp. 741–763.

Sola, P., Ochieng, C. and Yila, J. (2016) 'Links between energy access and food security in sub-Saharan Africa: An exploratory review', *Food Security*, vol. 8, pp. 635–642.

Subedi, M., Matthews, R., Pogson, M., Abegas, A., Balana, B., Oyesiku-Blakemore, J. and Smith, J. (2014) 'Can biogas digesters help to reduce deforestation in Africa?', *Biomass and Bioenergy*, vol. 70, pp. 87–98.

Surendra, K.C., Takara, D., Hashimoto, A.G. and Khanal, S.K. (2014) 'Biogas as a sustainable energy source for developing countries: Opportunities and challenges', *Renewable and Sustainable Energy Reviews*, vol. 31, pp. 846–859.

Tenenbaum, B., Greacen, C., Siyambalapitiya, T. and Knuckles, J. (2014) 'From the bottom up: How small power producers and mini-grids can deliver electrification and renewable energy in Africa energy and mining'. Washington, DC: World Bank. Available at: http://documents.worldbank.org/curated/en/507561468202200152/pdf/From-the-bottom-up-how-small-power-producers-and-mini-grids-can-deliver-electrification-and-renewable-energy-in-Africa.pdf

ter Heedge, F. and Sonder, K. (2007) 'Domestic biogas in Africa: A first assessment of the potential and need', draft/discussion paper. Available at: www.snv.org/public/cms/sites/default/files/explore/download/20070520_biogas_potential_and_need_in_africa.pdf

Tiepelt, M. (2015) 'Energy access', paper presented at South African International Renewable Energy Conference. Available at: www.sairec.org.za/wp-content/uploads/2015/10/-Tiepelt-SAIREC-Conference.pdf

UN (2018) 'Sustainable Energy "golden thread" linking 2030 agenda with pledge to leave no one behind, especially rural women'. Available at: www.un.org/press/en/2018/dsgsm1202.doc.htm

WBA (World Bioenergy Association) (2013) 'Biogas: An important renewable energy source', WBA fact sheet, World Bioenergy Association, Sweden. Available at: www.worldbioenergy.org/sites/default/files/wfm/Factsheet_Biogas.pdf

11 The need for stakeholder engagement and participative governance to promote bioeconomy

Divina Gracia P. Rodriguez and Anne Strøm Prestvik

Introduction

Bioeconomy has been gaining considerable attention in recent years due to the challenges related to climate change, environmental degradation, loss of biodiversity and increasing depletion of fossil fuel resources. The definition of bioeconomy links two main ideas, which focus on the use of renewable biological resources, and the use of biotechnology in production processes (Bugge et al., 2016). The bioeconomy can be defined as 'an economy where the basic building blocks for materials, chemicals, and energy are derived from renewable biological resources' (McCormick and Kautto, 2013). Its objective is to change how society produces, processes and recovers biological feedstocks in keeping with the aims of cleaner production, reduced waste and increased resource efficiencies (Devaney and Henchion, 2018). According to some, it may be a promising pathway to achieve several of the United Nation's Sustainable Development Goals (SDGs) and the Paris Agreement on climate commitments (Heimann, 2018).

The Bioeconomy Strategy of the European Commission (EC) is based on three key pillars: (1) investing in research, innovation and skills; (2) market development and enhanced competitiveness of bioeconomy sectors; and (3) a stronger policy coordination and engagement with stakeholders (EC, 2018). This strategy therefore requires:

1 *science* that relates to transformation of conventional production patterns and products into bio-based processes and products,
2 adaptive and inclusive *policies and regulations* to support the supply and demand for innovative bio-based products and materials;
3 changes in *stakeholder practices, behaviour and perceptions* about the need for the development of the bioeconomy and environmental protection.

Without these pre-conditions, attempts to promote bioeconomy might not provide the expected outcomes or could even result in the implemented processes failing (Sanz-Hernandez et al., 2019). However, in practice these three requirements typically operate in isolation of one another.

Stakeholder engagement practices for participative governance have been rare in the bioeconomy development process so far (Davies et al., 2016), despite the wide recognition of the significance of stakeholder engagement in such a process. For instance, the initiative for many bio-economy projects is rarely taken jointly by partners from research, policy and practice. This chapter therefore focuses on four main questions:

1 why there is a need for stakeholder engagement and participative governance;
2 who the relevant stakeholders are in bioeconomy development;
3 how stakeholders should be engaged;
4 what challenges are associated in the stakeholder engagement process.

It draws experiences from the EU and other regions where (formal and informal) stakeholder engagement and participative governance are seen as necessary factors for a successful bioeconomy development.

The need for stakeholder engagement and participative governance

While science is one important part of developing the bioeconomy, another part is to understand and guide the social change necessary for the implementation (Knierim et al., 2018). The social sciences often see the development of the bioeconomy as a political project, the solution to a 'wicked problem', or our era's great social transformation, all of which requires the involvement of actors from academia and outside academia into the research process. This is to integrate scientific and practical knowledge, reconcile values and preferences, as well as create ownership for problems and solution options in order to achieve the desired outcomes.

Stakeholder engagement

Stakeholder engagement can be defined as a process to actively involve stakeholders throughout a decision process by soliciting their knowledge, experience, judgement and values, 'for the dual purposes of creating a shared understanding; making relevant, transparent and effective decisions' (Deverka et al., 2012). In research, stakeholder engagement is a valuable means to provide societal perspectives concerning the policy, legal, ethical, and other impacts of scientific and technological research (PytlikZillig and Tomkins, 2011).

Most of the research on bioeconomy was developed from the suppliers' angle, i.e. the scientists in the lab, the extension agent in the field, and the bio-technology company in the marketplace, without end user consultations – the farmers who adopt the technologies, the consumers who demand new goods and services, or the communities who act collectively to improve their livelihoods and protect the environment. Policy-makers often use research findings, not to inform decisions but rather to back up decisions already made (Amara et al., 2004). As a result, there is a wide range of biophysical and socio-economic perspectives in which bioeconomy is developing that may hamper sustainable development.

An overview of history and typologies of stakeholder engagement

Many scholars have argued that engaging stakeholders in the research and policy-making process is a key element towards a sustainable bioeconomy. There are different typologies that can be used to select the methods that are likely to be most appropriate to the decision-making context, considering the appropriate level of engagement, type of participants and objectives of participation. Below is a brief history of the typologies of stakeholder engagement.

The approaches to stakeholder participation have progressed through a series of recognizable phases (Reed, 2008): (1) awareness raising in the late 1960s (van Tatenhove and Leroy, 2003); (2) incorporating local perspectives in data collection and planning in the 1970s (Pretty, 1995); (3) the development of techniques that recognized local knowledge 'put the last first', e.g. participatory approaches in agricultural research in the 1980s (Chambers, 1983); (4) increasing use of participation to enhance sustainable agriculture and natural resource management of the 1990s (e.g. UNCED, 1992); (5) the subsequent critique of participation and disillusionment over its limitations and failings (e.g. Cooke and Kothari, 2001); and (6) 'post-participation' consensus over best practice, improving participatory processes through new tools and methods (e.g. Hickey and Mohan, 2005).

There are many typologies to engage stakeholders typically seen in environmental management. The typologies of stakeholder engagement started in the 1960s with the 'ladder of participation' (Arstein, 1969) that recognizes categories ranging from passive dissemination of information, to consultation, to citizen control (Luyet et al., 2012). Numerous alternative terms have been suggested for the different rungs of this ladder. For example, Bigg (1989) described the level of engagement as a relationship that can be 'contractual', 'consultative', 'collaborative', and 'collegiate'. Pretty's (1995) 'participation scale' spans from manipulative and passive participation to interactive participation and self-mobilization. Today, many have considered the ladder of participation obsolete and debatable because it considered participation as an end rather than as means (Wehn and Evers, 2014). Nevertheless, many practitioners and decision-makers still use it as their point of reference.

Alternatively, Richards et al. (2004) suggested methods appropriate to the different levels of engagement on a 'wheel of participation' to emphasize the legitimacy of different degrees of engagement for different purposes and contexts (Davidson, 1998). Rowe and Frewer (2000) then based the typology on a wide range of methods that can be used to communicate (e.g. newsletters), consult (e.g. focus group discussions) or participate (e.g. meetings) with stakeholders. There are also typologies that distinguished between participation that is normative and/or pragmatic (Beierle 2002). Other typologies are based on the objectives for which participation/ engagement is used (Tippett et al., 2007), e.g. 'research-driven' versus 'development-driven' participation (Okali et al., 1994). Fung (2006), meanwhile, describes various approaches to participation through a 'democracy cube' that is based on axes of authority and power, types of participants, and communication and decision mode.

Instrumental and normative motivations for stakeholder engagement

Two main theories for why stakeholders should be engaged in the bioeconomy development and implementation are discussed here: the *instrumental and normative approaches* (Figure 11.1).

The instrumental approach sees stakeholder engagement as a process or a tool to achieve a certain goal, e.g. to make sure new technologies are adopted (Gerdes et al., 2018). This approach is sometimes split into two parts: (1) building trust and good relationships and reduction of (potential) conflict; and (2) the substantive approach (Marries and Rose, 2010),

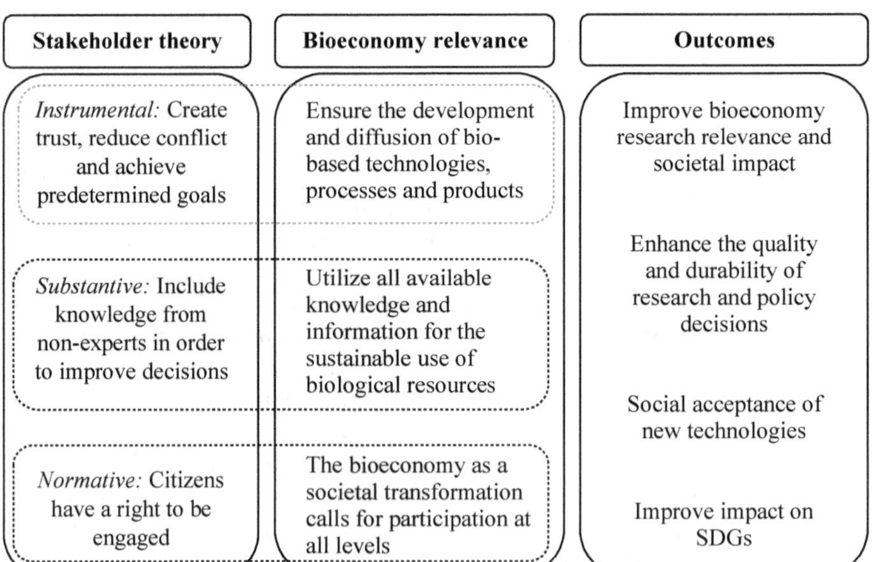

Figure 11.1 Stakeholder theory and the relevance for bioeconomy outcomes.

which is the collection of knowledge from non-experts to make more informed and better decisions. Building trust in the new bio-based technologies and good relationships with its users will foster social acceptance and implementation of new technologies and products among target groups. Stakeholders will have sense of ownership of the decisions that is expected 'to provide a higher degree of sustainable and innovative outcomes' (Heinelt, 2002). The integration of local knowledge and expertise into all aspects of bioeconomy strategies may prove to be both valuable and necessary for the successful implementation of the bioeconomy. This is not only true for science, but also for the development of targeted and efficient policies which needs input from a wide range of stakeholders (Devaney et al., 2017).

The normative approach, meanwhile, is where citizens have the 'right' to be involved in all the stages of decision-making, as the transition to bioeconomy requires societal transformation (Zeug et al., 2019) and maximum participation. It is the moral duty of decision-makers to involve citizens in decisions that affect their lives, given the scope of the political challenge in bioeconomy and attaining the SDG. A normative approach to stakeholder engagement will often lead to instrumental outcomes (Reed et al., 2009), as this justifies the engagement. It leads to the development of trust and understanding, exchange of knowledge, and builds a platform for working towards mutual goals and co-creation. Hence, stakeholder engagement strengthens the science-policy-stakeholders links as it offers an opportunity to examine and establish optimal pathways that may help achieve sustainable bioeconomy.

In the EU, the rationale for stakeholder engagement is from an instrumental perspective by improving the levels of stakeholders' trust through the provision of information. For instance, the stakeholder engagement through the EU public-private-partnership (PPP) with the Bio-Based Industries Consortium (BIC) has been instrumental in the development and deployment of new bio-based value chains based on the use of renewable resources including waste (EC 2017a). In 2012, there was a stakeholder consultation in the development of the EU Bioeconomy Strategy, which shows elements of the normative perspective, addressing the right of the stakeholders to provide inputs.

Participative governance

The bioeconomy approach is governed by a network of actors from different sectors who have partly aligned and partly conflicting interests. Implementing a sustainable bioeconomy pathway therefore depends not only on the technological feasibility and economic viability of selected production pathways but also on the management and engagement of a multitude of stakeholders (Langeveld et al., 2010) to ensure the sustainable allocation of resources across food, feed, fuel and fibre needs (Lewandoski, 2015). There are trade-offs and decisions to be made on the design and development of the bioeconomy, involving several sectors and end users whose interests conflict (McCormick and Kautto, 2013). For example, biofuel production may threaten food security because it competes with food production for land and even with some crops (e.g. wheat, corn, soybeans and sugar cane) and may contribute to the rise of food prices (Chakravorty et al., 2015). As new bioeconomy pathways emerge, the need for diverse stakeholder groups will become even more relevant in decision making. Figure 11.2 displays a conceptual framework that can be used to analyse the governance in bioeconomy.

The interaction of different actors in the bioeconomy leads to various governance challenges according to Birner (2018): (1) *political economy challenges*, i.e. the government themselves are subject to variety of forces, e.g. lobbying by industry groups, which may not necessarily be in favour of the bioeconomy; (2) *coordination*, i.e. fostering the collaboration among different stakeholders; and (3) *global bioeconomy governance*, i.e. reconciling different global concerns, such as food security and increasing use of biomass. These challenges may require participatory governance to improve bioeconomy outcomes. Participative governance can be defined as the 'processes and structures of public decision making that engage actors from the private sector, civil society, and/or the public at large, with varying degrees of communication, collaboration, and delegation of decision power to participants' (Newig et al., 2018). It will require bringing multi-stakeholders together to negotiate and compromise in reaching a consensus as to the optimally tangible solution that satisfies all their collective interests and needs (Fischer, 2012), taking into consideration the following: (1) more equal distribution of political power; (2) a fairer distribution of resources; (3) the decentralization of decision-making processes;

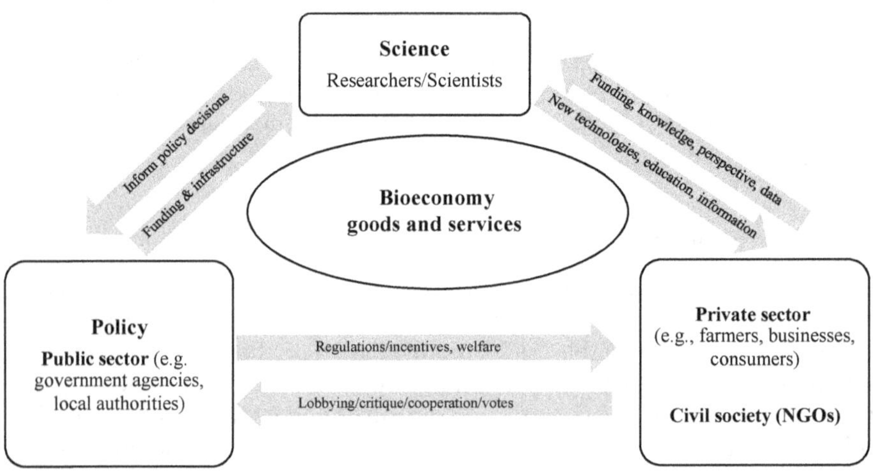

Figure 11.2 Stakeholder and governance of bieconomy.

(4) the development of a wide and transparent exchange of knowledge and information; (5) the establishment of collaborative partnerships; (6) an emphasis on inter-institutional dialogues; and (7) greater accountability.

The relevant stakeholders in bioeconomy

Stakeholders may be any actor – institution, group or individual – who fulfils the following conditions:

1. has the power/ability to make a real change towards the implementation of the bioeconomy (instrumental), e.g. policy-makers;
2. holds valuable information or knowledge for the issue at hand (substantive), e.g. scientists;
3. is directly or indirectly affected by the planning and implementation of the bioeconomy (normative), e.g. farmers and citizens.

As depicted in Figure 11.2, stakeholders can include:

1 *research/scientific organizations* who are believed to be key in the development of the knowledge based bioeconomy, not just in developing and sharing knowledge and enabling innovation processes, but to evaluate and monitor the sustainable bioeconomy development;

2 *private sector/organizations mainly include*

  a. *bio-based industries and small and medium-sized enterprises*, which produce mainly the bioeconomy products and services. They can benefit from government policies (e.g. incentives) and research on bioeconomy (e.g. new technologies, information, education).

  b. *farmers and farmer organizations* as commodity producers who will have a basic role in the bioeconomy as technology adopters, partners in innovation, and suppliers of biomass for the bioeconomy industry. The agricultural sector also provides public goods like agricultural landscapes and other ecosystem services (Schmid et al., 2012).

  c. *consumers, citizens and voters* – their acceptance of and trust in bioeconomy goods and services are important determinants of their success in commercialization.

3 *public sector/governmental institutions*: Policy-makers influence bioeconomy development and implementation directly and indirectly through strategies and policies that address the respective bioeconomy sectors. It is also both the duty of science and policy-makers to engage the public in their respective activities/decision-makings through supporting research and development, providing incentives for investing in the bioeconomy (Philp 2018), and creating coherent policy, including regulation and law (Dietz et al., 2018); and

4 *civil society organizations* (CSOs), also referred to as the non-governmental organizations (NGOs) or 'third sector', typically pursue public interests such as environmental protection. An example is the Biofuelwatch, a CSO mainly based in the UK and the US, that particularly advocates around the negative impacts of large-scale industrial bioenergy (Biofuelwatch.org.uk, 2019).

Involving stakeholders in planning and implementing bioeconomy strategies

Bioeconomy is conceptualized in different ways, associated with different expectations (Meyer, 2017). Accordingly, different models of stakeholder engagement are being adopted in the bioeconomy development globally in different regions.

Types of stakeholder engagement

A typology of stakeholder engagement has two dimensions. First, stakeholder engagement varies in mode, from one-way information in one end to co-production in the other as presented in Figure 11.3 (Gerdes et al., 2018; Reed et al., 2018). In the left panel, this dimension includes approaches where the only goal is to inform or educate stakeholders, to more consultative approaches where stakeholders are expressing their opinions. In the

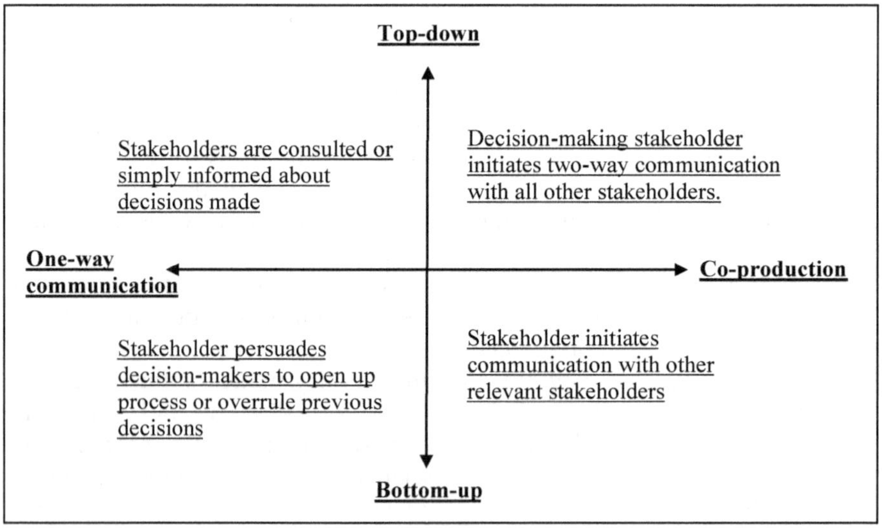

Figure 11.3 Types of stakeholder engagement.

Source: adapted from Reed et al. (2018); Gerdes et al. (2018).

right panel, this includes participatory approaches where there is dialogue and deliberation and where stakeholders can influence decision-making.

The second dimension relating to who initiates and leads the engagement process determines whether it is 'top-down' or 'bottom-up'. Top-down means the process is initiated by the stakeholder who has the power to make decisions, e.g. national government, as a deliberate process of strategy formation. Bottom-up happens when stakeholders, e.g. industry, take the initiative to influence decision-making. This process is usually more emergent, building on existing policy objectives and instruments that over time may lead to a more holistic strategy. The top-down/bottom-up distinction is difficult to determine in practice. On the same note, a stakeholder engagement process may be both top-down and bottom-up simultaneously. Below are four examples showing different modes of stakeholder engagement.

The *top-down one-way communication and/or consultation* represents processes where stakeholders are simply informed and/or consulted about the decision or decision-making processes, but the decisions are still made and implemented by those who hold power. The EU Water Framework Directive is an example of this approach, even though the intent was to move stakeholder engagement towards a more participatory mode (Box 11.1).

Box 11.1 Stakeholder engagement in the European Water Framework Directive

The European Water Framework Directive (WFD) was adopted in 2000 after growing concerns about water management and a need for common European water policy. The purpose of the WFD was to establish a framework for the protection of European waters for Member States to reach 'good status' objectives for water bodies throughout the EU (Voulvoulis et al., 2017).

The WFD can be categorized as a top-down consultation process. Even though Article 14 explicitly called for public participation, the drafting and implementation of the river basin management plans remained with the statutory bodies in each country. Fifteen years since it was adopted, however, the WFD still has not delivered its main objective of non-deterioration of water status and the achievement of good status for all EU waters (EC, 2012b).

An analysis of WFD implementation in EU Member States showed an increase in stakeholder engagement, dependent on existing institutional and procedural arrangements (Jager et al., 2016). In some cases, stakeholder participation was institutionalized through the formation of advisory groups. To achieve its targets, WFD needs a more complex, multilevel system of governance, including social learning processes, 'ownership' of environmental problems and public education (Woods, 2008).

Source: Adapted from Voulvoulis et al. (2017); Jager et al. (2016); Woods (2008).

Top-down deliberation and co-production occur when the power-holding stakeholder initiates two-way communication with other stakeholders and/or the general public. Co-productive processes involve stakeholder engagement at all stages and include deliberation. The decisions or other outcomes are owned by both the decision-maker and stakeholders. An example of this approach is the multi-actor approach adopted by many research projects (Box 11.2), and now a requirement by many research programmes and funding agents, e.g. several topics in the EU Horizon 2020 (EC, 2017b).

The 'Biomass Town' Project is a good example of bottom-up and co-production approach, focusing on the establishment of a comprehensive biomass utilization system to revitalize rural areas (MAFF, 2012). The plan was formulated through stakeholder consensus and implemented by the stakeholders who designed it. The stakeholders are engaged through

Box 11.2 Multi-actor approach in stakeholder engagement

Innovation is a necessary precondition for the success of bioeconomy. The European Innovation Partnerships-Agri is one of the five innovation partnerships launched by the EU. It focuses on agriculture and forestry and is bridging research and practice by bringing together innovation actors (e.g. researchers and farmers) (EC, 2014). The overarching model is called the 'interactive innovation model', which is a collaboration among actors that leads to co-creation and diffusion of innovations ready to be put into practice. Knowledge that leads to innovation does not necessarily need to come from science or research, it can come from farmers, NGOs, and small and medium-sized enterprises.

Innovation in research projects supported by the Horizon 2020 is ensured through the 'multi-actor' approach and is a requirement for several topics (e.g. Sustainable Intensification Africa) (EC, 2017b). The aim of the approach is to make innovation more demand-driven by involving end-users and other stakeholders in the project activities from the planning stage to the implementation and dissemination (ibid.).

The different stages are: (1) *co-analysis*: includes diagnosing problems, constraints, and solutions together with all stakeholders involved; (2) *co-design*: implies that the understanding of a problem and/or solution can be improved if local stakeholders/experts are actively involved and their knowledge is integrated when identifying and prioritizing solutions. The co-design process looks beyond overcoming immediate constraints and will outline pathways that are sustainable in the long run. All stakeholders in the project are expected to work together in the validation of technologies and joint assessment of results.

Source: Adapted from EC (2014, 2017b).

institutional arrangements (e.g. cooperatives and regional clusters). The Biomass Town has been implemented in Japan, Thailand, Vietnam, Malaysia and Indonesia. In Thailand, the goal was to implement biomass utilization technologies in the villages so that villagers can use these products in their households, farmlands, and livestock farms. In the project, strong collaboration was made possible through the support from various government agencies both domestically and overseas (Tantiwatthanaphanich and Zou, 2016).

Examples of bottom-up, one-way communication are how civil society organizations (CSOs) are attempting to influence decisions regarding bioeconomy development. In 2018, a coalition of 120 CSOs from 40 countries presented an open letter to the BioFuture Platform (a 20-country attempt to promote and advance bioeconomy (Biofutureplatform.org, 2018)), questioning the sustainability of a rapidly growing bioeconomy (Quaranda, 2018).

Types of stakeholder engagement: what works and what does not?

As we recognize the importance of stakeholder engagement and participative governance in bioeconomy transformation, it is also imperative to recognize the need to evaluate both the processes and outcomes of different approaches to stakeholder engagement and participative governance. Even though the more participative modes of stakeholder engagement are associated with democratic values, one-way communication may be appropriate in some situations, depending on the objective. There are five factors that can explain much of the variation in stakeholder engagement (Reed et al., 2018) that are applicable to bioeconomy planning and implementation strategies. These are:

1 the *context*, which includes the specific social, economic, political, technological, and environmental settings;
2 the *design* of the engagement process, in which stakeholders' interests, values and the expected outcomes are taken into consideration;
3 the *power* – the existing power relations may influence how stakeholders interact, who are included in the process and to what degree they are able to participate and influence the decision-making. Figure 11.4 presents the levels of engagement that also reflect the level of power/influence stakeholders have in the decision-making processes (IAP2, 2000). As the level of participation increases, the concerned stakeholders' power and influence on decision also increase.
4 *scalar fit* refers to the temporal and spatial conditions of jurisdictions and institutions;
5 the *capacity and availability of resources* of both the stakeholder leading the engagement process (e.g. research institution) and the

Level of participation	Characteristics of the level
Empower	Stakeholders take responsibility for designing and delivering services themselves.
Collaborate	Stakeholders go through shared decision-making, including the development of alternative methods and the identification of preferred solutions or outcomes
Involve	Stakeholders work together throughout the entire project life cycle but the decision-making responsibility rests with the proponent or authority.
Consult	Obtaining feedback from and provide adequate information to stakeholders to inform decision-making
Inform	Stakeholders are provided with information to enable them to understand problems, alternatives, opportunities and solutions related to bioeconomy.

Figure 11.4 Five levels of stakeholder engagement.

Source: IAP2 (2000).

stakeholders being asked to participate in it (e.g. farmers). Factors that can affect stakeholders' capacity include time, money, training, and experience. In Finland and Germany, for example, one challenge in engaging a range of organizational stakeholders in bioeconomy strategies is that co-funding requirements can limit the scope of some entities (e.g. SMEs) wishing to obtain public funding (Griestop 2018). This can be overcome by providing funding mechanisms to engage stakeholders as demonstrated by EU H2020 programme.

Stakeholder engagement and status of participative governance in bioeconomy strategies

Attempts are already being made to engage not just scientists and policymakers in the bioeconomy planning and implementation, but also other relevant stakeholders. At the EU level, in 2012, the EC produced a combined strategy and action plan entitled 'Innovating for Sustainable Growth: A Bioeconomy for Europe' (EC, 2012a) in consultation with stakeholders. The strategy recognizes the importance of stakeholders in knowledge transfer and implementation. To improve policy interactions, the Bioeconomy Panel (later called the Bioeconomy Stakeholder Panel in 2016), consisting of representatives from the EC and Member States and stakeholders from the private sector, research, and civil society was also set up. This has enhanced and integrated stakeholder engagement, making policy development more socially inclusive and visible in the EU (EC, 2017b).

At the national level, many countries now have their national bioeconomy or similar strategies, mainly focusing on research, innovation, and development. Table 11.1 provides an overview of the different stakeholder

Table 11.1 Examples of stakeholder engagement and participatory governance in bioeconomy strategies

	Context	*Design/method*	*Power/ level of influence*	*scale*	*Source OF FUNDING*
Developed countries					
Belgium –Flanders	Research, innovation, and policy	Stakeholder consultation	Consult	Regional level, 2014–2030	Mix of public and private funding
Finland	Renewable resources	Bioeconomy Panel, forum	Consult	National level, 2015–2025	Public funding
France	Innovation/Industrial	Bioeconomy advisory council, PPP, clusters,	Consult	National level, 2014–2020/2050	Mix of public and private funding
Germany	Research, development and policy	IWG, German Bioeconomy Council	Consult	National level, Research strategy (2010–2016), Policy strategy (2010–n/a)	Public funding
Italy	Innovation/Industrial	IWG, PPP, clusters	Empower	National level, 2015–	Mix of public and private funding
Norway	Research and innovation	Clusters, PPP, networking activities	Consult	National level, 2012–2022	Mix of public and private funding
Canada	Innovation/Forest	IWG, PPP, clusters, networks	Collaborate	Regional level, 2013–2018	Mix of public and private funding
Japan	Research and innovation	IWG, Biomass Policy Council	Consult	National level, 2010–2020	Public funding
USA	Research and innovation	IWG, PPP, clusters, networks, multi-stakeholder dialogue	Consult	Federal, no specified time horizon	Public funding

continued

Table 11.1 continued

	Context	*Design/method*	*Power/ level of influence*	*scale*	*Source OF FUNDING*
Emerging economies					
Brazil	Research and innovation	IWG, PPP, forum	Consult	National level, 2004, 2007	Mix of public and private funding
Russia	Research and innovation	IWG, PPP, clusters, platforms	Collaborate	National, 2012–2020	Mix of public and private funding, international capital markets
India	Research, innovation, and development	Public consultation	Consult	National level, 2007, 2014	Public funding
China	Research, innovation, and development	The State Council, PPP, clusters, innovation networks and hubs	Inform	National level, 2016–2020	Mix of public and private funding
Developing countries					
Malaysia	Research, development and policy	PPP, multi-stakeholder platform	Consult	National level, 2013	Public funding and mix of public and private funding
Thailand	Innovation/Industrial	IWG, PPP, innovation hubs	Empower	National level, 2012–2021	Mix of public and private funding
South Africa	Research, development and innovation	IWG	Consult	National level, 2014	Public funding
Kenya	Research and innovation/ bio-diesel/bioethanol	Advisory Body, communication campaigns	Consult	National level, 2008–2012 (Biodiesel), 2009–2012 (Bioethanol)	Public funding

engagement approaches in national and regional bioeconomy strategies. And Table 11.2 presents the various stakeholders involved in the development of these strategies by different countries.

National bioeconomy strategies in emerging economies like Brazil, China, India, Malaysia, Russia and South Africa were largely prepared with a top-down approach. In countries such as Finland, Germany and Norway, public consultations were included as part of the bioeconomy strategy preparations. The strategy in Germany, for example, was developed jointly by several ministries coordinated and led by the Federal Ministry for Food and Agriculture. With its explicit focus on policy coordination, inter-ministerial collaboration and coherence were addressed by institutionalizing an Inter-Ministerial Working Group on Bioeconomy (IWG). The strategy included cross-sectoral activities, such as transdisciplinary research, facilitating implementation of new innovations and technologies through actions of stakeholders from research, private, and public sectors. This facilitates stakeholders to build strategic alliances along the whole production chain of bioeconomy and increases the chances of receiving federal funds (Overbeek et al., 2016). Moreover, the strategy was discussed with representatives from agriculture and the business sector, universities, NGOs and selected regional authorities (Davies et al., 2016).

The German Bioeconomy Council, composed of representatives from research and the private sector, was also created and played an important role in the strategy development. It is recognized as being increasingly transparent, inclusive and open for public debate (ibid.). Over time, its mandate has broadened to include monitoring and evaluation of the bioeconomy (Imbert et al., 2017). This is to measure progress in attaining the ambitions and targets set in bioeconomy strategies. Experience from Germany can be useful to other countries that are trying to develop bioeconomy.

Conversely, France, Italy, Canada and Belgium whose bioeconomy strategies vary by region, have adopted bottom-up and collaboration strategies to leverage existing PPPs. They mainly focus on their industrial and technological leadership to develop and implement bioeconomy strategies. Co-funding by the private sector therefore is important in the bioeconomy implementation. For years, Italian bioeconomy development, for instance, has been led by stakeholders in the chemical industry (Ronzon et al., 2015).

To ensure coordination, identification of common strategies and cooperation on bioeconomy-related projects, these countries widely used policy instruments such as clusters, hubs or public-private partnerships (PPPs). Clusters are groups of stakeholders, which specialize in a particular technology to build an 'industrial ecosystem' (Philp and Winickoff, 2017). Clusters are instrumental to leverage the participation of SMEs in the bioeconomy. In 2005, for example, the cluster policy introduced by the

Table 11.2 Stakeholders involved in bioeconomy strategy development

Country	*Research/ Science*	*Private sector*			*Public sector*	*Civil society*	*Reference*
		Industries/ businesses	*Farmers*	*Consumers/ citizens*			
Developed countries							
Belgium: Flanders	✓	✓	–	–	✓	✓	http://www.vlaanderen.be (2014)
Finland	✓	✓	✓	✓	✓	✓	Davies et al. 2016
France	✓	✓	–	–	✓	–	German Bioeconomy Council (2018)
Germany	✓	✓	✓	–	✓	–	German Bioeconomy Council (2018)
Italy	✓	✓	✓	✓	✓	–	Bonaretti et.al. (2017)
Norway	✓	✓	–	–	✓	–	Bardalen (2016)
Non-EU							
Japan	✓	–	–	–	✓	–	MAFF (2012)
Canada	✓	✓	✓	✓	✓	–	Government of Canada (2019)
USA	✓	–	–	–	✓	–	German Bioeconomy Council (2018)
Emerging economies							
Brazil	✓	–	–	–	✓	–	MSTI (2016)
Russia	✓	✓	–	–	✓	–	BIO2020 (2012)
India	✓	–	–	–	✓	–	Republic of India (2014)
China	✓	–	–	–	✓	–	NDRC (2016)
Developing countries							
Kenya	✓	✓	–	–	✓	–	Republic of Kenya (2006)
South Africa	✓	–	–	–	✓	–	Republic of South Africa (2013)
Malaysia	✓	✓	–	–	✓	–	MOSTI (2014)
Thailand	✓	✓	–	–	✓	–	Sirisaengtaksin (2017)

French government to boost bioeconomy innovation and develop innovation platforms allowed industrial demonstration and feasibility at the Technology Readiness Level (TRL) to jump from 5 to 9 (Stadler and Chauvet, 2017). In Thailand, 'Biopolis', a biotechnology cluster, was initiated following the successful example in Singapore. The main aim of Biopolis is to strengthen the national and international research bioeconomy networks.

Public-private partnerships can also: (1) foster collaboration of stakeholders to encourage successful market integration of bioeconomy goods and services; (2) encourage the establishment of innovation hubs/centres to ensure continuous learning from best practices; and (3) provide substantial support for innovative business clusters and networks. The PPP model is now considered a successful way of promoting research and innovation, reducing the risk for investment in research and works as a catalyst for private initiatives in the EU bioeconomy (Mengal et al., 2017).

In both top-down and bottom-up approaches, research, private and public sectors are often represented in strategy development, especially in the developed countries. While the private sector is usually represented by industries/business enterprises, the involvement of farmers and consumers/citizen is limited. CSOs have played a minimal role in the development and implementation of bioeconomy strategies. One exception is Finland, where CSOs were involved in the Stakeholder Panel. In most of the emerging and developing countries, the involvement of farmers, citizens/consumers and civil society is often missing. Nevertheless, a broad stakeholder involvement in implementation is the major objective of most national bioeconomy strategies (Overbeek et al., 2016).

Challenges in stakeholder engagement

Stakeholder engagement and participative governance in bioeconomy are not without challenges and limitations. So far, CSOs have played a marginal role in the process and implementation of bioeconomy strategies, and there is a concern that critical voices have not been heard in the high-level debate (Meyer, 2017). Even though most consultation processes have been open to the general public, this may not be enough to engage the civil society and other members of the private sector (i.e. farmers and consumers/citizens).

Information asymmetries and power imbalances

While most bioeconomy strategies recognize the importance of stakeholder engagement in the bioeconomy development and implementation, this does not mean that stakeholder engagement always leads to tangible environmental and social benefits. It can fail to deliver expected outcomes (e.g. Staddon et al., 2015) and/or can be counterproductive in achieving the democratic goals often ascribed to it (e.g. Tewdwr-Jones and Allmendinger, 1998). Stakeholder engagement can result in:

1 *elite capture* intended as the capacity of 'elites' to manoeuvre benefits from bioeconomy programmes and projects for themselves. For example, developed countries such as Italy promote bio-based applications in the chemical industry sectors. A few powerful and resourceful groups of stakeholders can use their influence to make patent rights very restrictive and licensing procedures very lengthy and costly. Hence, the associated benefits accrue only to the elite segment of the population.
2 *information asymmetries* that occur when one or more stakeholders have more or better information than the others (Moe, 1984). Those who have more access to information and resources are more likely to influence policy directions that do not necessarily contribute to the public good.
3 *collective action problems* which happen when all stakeholders would be better off cooperating but fail to do so because of conflicting interests between stakeholders that discourage joint action (Ostrom, 1990);
4 *principal-agent problems* occur when, for example, BIC in the EU can raise challenges in terms of lobbying in which the consortium's representative (the agent) is allowed to make decisions on behalf of the whole consortium (the principal). While formal engagement mechanisms like BIC and Biopolis do bring some advantages (e.g. Biopolis encompasses stakeholders invested in similar activities to exchange good practices and develop solutions and innovations together), they tend to look out for their own self-interests and can focus solely on pushing forward the agenda of their group.

Institutional fragmentation

Researchers, the private sector, and civil society must be linked more closely with multilateral policy processes and intergovernmental discussions (Ingrao et al., 2018). For this to happen, however, it is a necessary condition to bring different sectors in the government itself to work together. The EU Bioeconomy strategy from 2012 specifies the need for a cross-sectoral and cross-policy approach across the EU and beyond (EC, 2017b). While inter-ministerial bodies are set up, such as in Germany, the issue is sustainability of this collaboration since there is a wide range of political and societal issues involved.

Discontinuous participation

Maintaining stakeholders' continuous interest and active participation can be a challenge. While clusters and PPPs have proved to be successful in some countries in connecting stakeholders and increasing market opportunities, these instruments require a long-term strategy and support, in which the roles of public and private partners (as well as civil society) are essential. Stakeholder fatigue may also happen when stakeholder engagement initiatives have taken place in the past but have resulted in poor personal reward for stakeholders (Ruth, 2009).

Limitations on stakeholder engagement methods

The PPPs, such as the Biomass Town concept, are one way of speeding up developments in bioeconomy. Even if such bottom-up initiatives are effective in reaching their goals, the question remains whether such approaches are sustainable in the long term. When such partnerships are started with a significant amount of public funding, there is always a danger of developing a dependence on that funding. There is also the question whether such processes are inclusive, so that the outcomes also benefit marginalized groups and the environment.

Conclusion

Stakeholder engagement and participative governance are key in strategy development, planning and implementation of bioeconomy, as demonstrated in many national and regional bioeconomy strategies. They are also a necessary element to strengthen the science-policy-stakeholder linkage, which is essential for successful implementation of the bioeconomy. Engaging stakeholders may help to achieve sustainable bioeconomy by improving: the bioeconomy research relevance and societal impact through identification of socially relevant problems (Slunge et al., 2017); integrating the perspectives of a multitude of actors (Koontz and Thomas, 2006); enhancing the quality and durability of research and policy decisions; accessing and utilizing all available information and non-expert knowledge to make more informed and targeted policy decisions (Ruth, 2009); fostering social acceptance and implementation of new technologies and products among target groups through inclusive engagement process (Kastens and Newig, 2008); and linking the attainment of SDGs to the transition to bioeconomy (Zeug et al., 2019).

There are many forms of stakeholder engagement, from simply informing and educating stakeholders, to co-production, such as PPPs and transdisciplinary research. Sometimes the national government needs to take the lead, but there are also cases of bottom-up stakeholder engagement. Context, design, power, scalar fit and capacity and resource availability of stakeholders are factors that can explain the variation in outcomes from different types of stakeholder engagement. Stakeholder engagement does not necessarily yield positive and/or expected outcomes. It can result in well-known problems such as elite capture, information asymmetries, collective action problems and principal-agent problems. Many challenges remain such as biased representation of stakeholders or key stakeholders missing, discontinuous participation, institutional fragmentation, and limitations in stakeholder engagement methods.

Bioeconomy planning and development form a multidimensional and complex process involving several stakeholders. Therefore, stakeholder engagement processes should consider their interests, enable multiple stakeholders to respond and adapt to uncertainty and manage risks, in which changes in knowledge, technologies, and institutional environment are considered. While there are examples of effective stakeholder engagement, such as PPPs and industrial clusters, active stakeholder participation is still inadequate in the development of most country strategies and many initiatives. Regardless of the stakeholder engagement approach, national bioeconomy strategies are often centred on technical and industrial issues of bioeconomy development. Therefore, efforts to involve civil society beyond establishing information channels for one-way communication should be prioritized. The lack of a broad involvement of stakeholders may be a concern, especially when discussing questions such as sustainability, which needs a broad societal debate and acceptance (ibid.). Existing power structures and institutions will continue to influence the 'who and how' of stakeholder engagement, but new and innovative methods involving the civil society may break some of these patterns and ensure broader participation (Gerdes et al., 2018).

National bioeconomy stakeholder panels, consisting of representatives from all stakeholder sectors, may play an important role in creating public debate and engaging the civil society. In addition, the multi-actor approach in bioeconomy projects, now a requirement by many research and funding programmes, can facilitate different stakeholders to work together from the beginning, providing regular feedback on progress towards outputs and outcomes of the bioeconomy strategies and action plans. Stakeholder participants can evaluate the objectives and outputs of the project and/or policy themselves of whether they reflect the needs of all relevant stakeholders throughout the project and/or policy cycle (Phillipson and Liddon, 2007). Results of this evaluation can feed back into both theoretical knowledge and practical implementation of stakeholder engagement (Acompab et al., 2013; IPCC, 2014).

References

Acompab, D.A., Peng, B., Williams, S., Saniotis, A. and Walker, I.A. (2013) 'Engaging stake-holders in an adaptation process: Governance and institutional arrangements in heat-health policy development in Adelaide, Australia', *Mitigation and Adaptation Strategies for Global Change*, vol. 18, no. 7, pp. 1001–1018.

Amara, N., Ouimet, M. and Landry, R. (2004) 'New evidence on instrumental, conceptual, and symbolic utilization of university research in government agencies', *Science Communication*, vol. 26, pp. 75–106.

Arstein, A. (1969) 'A ladder of citizenship participation', *Journal of the American Institute of Planners*, vol. 26, pp. 216–233.

Bardalen, A. (2016) 'The Norwegian bioeconomy strategy: Structural changes and green shift in the economy'. Available at: www.norden.lv/Uploads/2016/08/26/1472194554_.pdf

Beierle, T.C. (2002) 'The quality of stakeholder-based decisions', *Risk Analysis*, vol. 22, pp. 739–749.

Bigg, S. (1989) 'Resource-poor farmer participation in research: A synthesis of experiences from nine national agricultural research systems', *OFCOR Comparative Study Paper*, vol. 3. The Hague, the Netherlands: International Service for National Agricultural Research.

BIO2020 (2012) 'Summary state coordination program for the development of biotechnology in the Russian Federation until 2020 "BIO 2020"'. Available at: http://bioeconomy.ru/upload/BIO2020%20(eng)%20-%20short.pdf

Biofuelwatch (2019) Biofuelwatch official website. Available at: www.biofuelwatch.org.uk/about/

Biofuture Platform (2018) Biofuture Platform official website. Available at: http://biofutureplatform.org/

Birner, R. (2018) 'Bioeconomy concepts', in I. Lewandowski (ed.) *Bioeconomy: Shaping the Transition to a Sustainable, Biobased Economy*. Cham: Springer International Publishing.

Bonaretti, P., Firpo, S., Tonci, C., Pasquali, O., Fava, F., Bianchi, L., Aleandri, R., et al. (2017) 'Bioeconomy in Italy: A unique opportunity to reconnect the economy, society and the environment'. Available at: https://ec.europa.eu/research/bioeconomy/pdf/bit_en_strategy_2019.pdf.

Bugge, M., Hansen, T. and Klitkou, A. (2016) 'What is the bioeconomy? A review of the literature', *Sustainability*, vol. 8, no. 7, p. 691.

Chambers, R. (1983) *Rural Development: Putting the Last First*. London: Longman.

Chakravorty, U., Hubert, M., Moreaux, M. and Nøstbakken, L. (2015) 'The long run impact of biofuels on food prices', CESifo Working Paper Series, No. 3876. Available at: https://ssrn.com/abstract=2104053

Cooke, B. and Kothari, U. (2001) *Participation: The New Tyranny?* London: Zed Books.

Davidson, S. (1998) 'Spinning the wheel of empowerment', *Planning*, vol. 1262, pp. 14–15.

Davies, S., Griestop, L., Vironen, H., Bachtler, J., Dozhdeva, V. and Michie, R. (2016) 'BioSTEP promoting stakeholder engagement and public awareness for a participative governance of the European bioeconomy: Case study of national bioeconomy strategies in Finland and Germany'. Available at: www.bio-step.eu/fileadmin/BioSTEP/Bio_documents/BioSTEP_D3.1_Case_studies_of_national_strategies.pdf

Devaney, L. and Henchion, M. (2018) 'Consensus, caveats and conditions: International learnings for bioeconomy development', *Journal of Cleaner Production*, vol. 174, pp. 1400–1411.

Devaney, L., Henchion, M. and Regan, Á. (2017) 'Good governance in the bioeconomy', *EuroChoices*, vol. 16, no. 2, pp. 41–46.

Deverka, P., Lavallee, D., Desai, P., Esmail, L., Ramsey, S., Veenstra, D. and Tunis, S. (2012) 'Stakeholder participation in comparative effectiveness research:

Defining a framework for effective engagement', *Journal of Comparative Effectiveness Research*, vol. 1, no. 2, pp. 181–194.

Dietz, T., Börner, B., Förster, J. and von Braun, J. (2018) 'Governance of the bioeconomy: A global comparative study of national bioeconomy strategies', *Sustainability*, vol. 10, p. 3190.

EC (European Commission) (2011) 'Bio-based economy for Europe: state of play and future potential: Part 1. Report on the European Commission's Public on-line consultation'. Available at: https://ec.europa.eu/research/consultations/bioeconomy/bio-based-economy-for-europe-part1.pdf

EC (European Commission) (2012a). 'Innovating for sustainable growth: A bioeconomy for Europe'. Available at: https://publications.europa.eu/en/publication-detail/-/publication/1f0d8515-8dc0-4435-ba53-9570e47dbd51

EC (European Commission) (2012b). 'Environment: Commission asks Italy to transpose water legislation correctly'. Available at: https://europa.eu/rapid/press-release_IP-12-292_en.htm?locale=en

EC (European Commission) (2014) 'Guidelines on programming for innovation and the implementation of the EIP for agricultural productivity and sustainability'. Available at: https://ec.europa.eu/eip/agriculture/sites/agri-eip/files/eip-guidelines-july-2014_en.pdf

EC (European Commission) (2016) 'Agricultural knowledge and innovation systems towards the future: A foresight paper'. Available at: https://ec.europa.eu/research/scar/pdf/akis-3_end_report.pdf

EC (European Commission) (2017a) 'Interim evaluation of the bio-based industries joint undertaking (2014–2016) operating under Horizon 2020'. Available at: https://ec.europa.eu/research/evaluations/pdf/bbi.pdf

EC (European Commission) (2017b) 'Horizon 2020 Work Programme 2018–2020. 9. Food security, sustainable agriculture and forestry, marine, maritime and inland water research and the bioeconomy'. Available at: https://ec.europa.eu/programmes/horizon2020/en/h2020-section/food-security-sustainable-agriculture-and-forestry-marine-maritime-and-inland-water

EC (European Commission) (2018) 'A sustainable bioeconomy for Europe: Strengthening the connection between economy, society and the environment'. Available at: https://ec.europa.eu/research/bioeconomy/pdf/ec_bioeconomy_strategy_2018.pdf.

Fischer, F. (2012) 'Participatory governance: From theory to practice', in D. Levi-Faur (ed.) *The Oxford Handbook of Governance*. Oxford: Oxford University Press.

Fung, A. (2006) 'Varieties of participation in complex governance', *Public Administration Review*, vol. 66, pp. 66–75.

Gerdes, H., Kiresiewa, Z., Beekman, V., Bianchini, C., Davies, S., Griestop, L., Janssen, R., et al. (2018) 'BioSTEP engaging stakeholders and citizens in the bioeconomy: Lessons learned from BioSTEP and recommendations for future research', Berlin, Germany: Ecologic Institute.

German Bioeconomy Council. (2018) 'Bioeconomy policy (part III): Synopsis of national strategies around the world'. Available at: https://biooekonomierat.de/fileadmin/international/Bioeconomy-Policy_Part-III.pdf

Government of Canada (2019). *Growing Forward* 2. Available at: www.agr.gc.ca/eng/about-us/key-departmental-initiatives/growing-forward-2/?id=1294780620963

Griestop, L. (2018) 'BioSTEP promoting stakeholder engagement and public awareness for a participative governance of the European bioeconomy: BioSTEP in a nutshell – overview and summary of project results'. Berlin, Germany: Ecologic Institute.

Heimann, T. (2018) 'Bioeconomy and SDGs: Does the bioeconomy support the achievement of the SDGs?', *Earth's Future*, vol. 7, no. 1, pp. 43–57.

Heinelt, H. (2002) 'Achieving sustainable and innovative policies through participatory governance in a multi-level context: theoretical issues', in H. Heinelt, P. Getimis, G. Kafkalas, R. Smith and E. Swyngedouw (eds) *Participatory Governance in Multi-Level Context: Concepts and Experience*. Opladen: Leske + Budrich.

Hickey, S. and Mohan, G. (2005) *Participation: From Tyranny to Transformation? Exploring New Approaches to Participation in Development*. London: Zed Books.

IAP2 (International Association of Public Participation) (2000) 'Public participation spectrum'. Available at: https://iap2canada.ca/Resources/Documents/0702-Foundations-Spectrum-MW-rev2%20(1).pdf

Imbert, E., Ladu, L., Morone, P. and Quitzow, R. (2017) 'Comparing policy strategies for a transition to a bioeconomy in Europe: The case of Italy and Germany', *Energy Research and Social Science*, vol. 33, pp. 70–81.

Ingrao, C., Bacenetti, J., Bezama, A., Blok, V., Goglio, P., Koukios, E.G. Lindner, M., et al. (2018) 'The potential roles of bio-economy in the transition to equitable, sustainable, post fossil-carbon societies: Findings from this virtual special issue', *Journal of Cleaner Production*, vol. 204, pp. 471–488.

IPCC (Intergovernmental Panel on Climate Change) (2014) *Fifth Assessment Report of the Intergovernmental Panel on Climate Change*. Cambridge: Cambridge University Press.

Jager, N., Challies, E., Kochskämper, E., Newig, J., Benson, D., Blackstock, K., Collins, K., et al. (2016) 'Transforming European water governance? Participation and river basin management under the EU Water Framework Directive in 13 Member States', *Water*, vol. 8, no. 4, p. 156.

Kastens, B. and Newig, J. (2008) 'Will participation foster the successful implementation of the WFD? The case of agricultural groundwater protection in northwest Germany', *Local Environment*, vol. 13, pp. 27–41.

Knierim, A., Laschewski, L. and Boyarintseva, O. (2018) 'Inter- and trans-disciplinarity in bioeconomy', in I. Lewandowski (ed.) *Bioeconomy*. Cham: Springer International Publishing.

Koontz, T. and Thomas, C. (2006) 'What do we know and need to know about the environmental outcomes of collaborative management?' *Public Administration Review*, vol. 66, pp. 111–121.

Langeveld, H., Meeusen, M. and Sanders, J. (eds) (2010) *The Biobased Economy: Biofuels, Materials and Chemicals in the Post-Oil Era*. London: Earthscan.

Lewandoski, I., (2015) 'Securing a sustainable biomass supply in a growing bioeconomy', *Global Food Security*, vol. 6, pp. 34–42.

Luyet, V., Schlaepfer, R., Parlange, M. and Buttler, A. (2012) 'A framework to implement stakeholder participation in environmental projects', *Journal of Environmental Management*, vol. 111, pp. 213–219.

MAFF (Ministry of Agriculture, Forestry and Fisheries, Japan), The Cabinet and National Biomass Policy Council. (2012) 'Reference 6. Biomass Policies and Assistance Measures in Japan'. Available at: www.maff.go.jp/e/pdf/reference6-8.pdf

Marries, C. and Rose, N. (2010) 'Open engagement: Exploring public participation in the biosciences', *PLoS Biology*, vol. 8, no. 11, e1000549.

McCormick, K. and Kautto, N. (2013) 'The bioeconomy in Europe: An overview', *Sustainability*, vol. 5, pp. 2589–2608.

Mengal, P., Zika, E., Wubbolts, M., Rúiz, A., Brigitta, D., Pieniadz, A. and Black, S. (2017) 'Bio-based industries joint undertaking: The catalyst for sustainable bio-based economic growth in Europe', *New Biotechnology*, vol. 40, pp. 31–39.

Meyer, R. (2017) 'Bioeconomy strategies: Contexts, visions, guiding implementation principles and resulting debates', *Sustainability*, vol. 9, p. 1031.

Ministry of Science, Technology and Innovation. (2016) 'Estratégia Nacional de Ciência, Tecnologia e Inovação (2016–2019)'. Available at: www.mcti.gov.br/documents/10179/1712401/Estrat%C3%A9gia+Nacional+de+Ci%C3%AAncia%2C%20Tecnologia+e+Inova%C3%A7%C3%A3o+2016-2019/0cfb61e1-1b84-4323-b136-8c3a5f2a4bb7

Ministry of Science, Technology and Innovation (MOSTI), Malaysia (2014) 'Bioeconomy transformation programme', *Annual Report*. Available at: www.biotechcorp.com.my/wp-content/uploads/2011/11/publications/V9_Booklet_version_FA.pdf

Moe, T. (1984) 'The new economics of organization', *American Journal of Political Science*, vol. 28, no. 11, pp. 793–777.

National Development and Reform Commission (NDRC). (2016) 'The 13th five-year plan for economic and social development of the People's Republic of China (2016–2020)'. Available at: http://en.ndrc.gov.cn/newsrelease/201612/P020161207645765233498.pdf

Newig, J., Challies, E., Jager, N., Kochskaemper, E. and Adzersen, A. (2018) 'The environmental performance of participatory and collaborative governance: A framework of causal mechanisms', *Policy Studies Journal*, vol. 46, no. 2, pp. 269–297.

Okali, C., Sumberg, J. and Farrington, J. (1994) *Farmer Participatory Research*. London: Intermediate Technology Publications.

Ostrom, E. (1990) *Governing the Commons: The Evolution of Institutions For Collective Action*. Cambridge: Cambridge University Press.

Overbeek, G., de Bakker, E., Beekman, V., Davies, S., Kiresiewa, A., Delbrück, S., Ribeiro, M.S., et al. (2016) 'BioSTEP promoting stakeholder engagement and public awareness for a participative governance of the European bioeconomy: Review of bioeconomy strategies at regional and national levels'. Available at: www.bio-step.eu/fileadmin/BioSTEP/Bio_documents/BioSTEP_D2.3_Review_of_strategies.pdf

Phillipson, J. and Liddon, A. (2007) 'Common knowledge? An exploration of knowledge transfer. Rural Economy and Land Use Programme', *Briefing Series*, no 6. Available at: www.relu.ac.uk/news/briefings/RELUBrief6%20Common%20Knowledge.pdf

Philp, J. (2018) 'The bioeconomy, the challenge of the century for policy makers', *New Biotechnology*, vol. 40, pp. 11–19.

Philp, J. and Winickoff, D.E. (2017) 'Clusters in industrial biotechnology and bioeconomy: The roles of the public sector', *Trends in Biotechnology*, vol. 35, no. 8, pp. 682–686.

Pretty, J.N. (1995) 'Participatory learning for sustainable agriculture', *World Development*, vol. 23, no. 8, pp. 1247–1263.

PytlikZillig, L. and Tomkins, A. (2011) 'Public engagement for informing science and technology policy: What do we know, what do we need to know, and how will we get there?', *Review of Policy Research*, vol. 28, no. 2, pp. 197–217.

Quaranda, S. (2018). 'Release: The dark side of the bioeconomy'. Available at: www.dogwoodalliance.org/2018/11/release-the-dark-side-of-the-bioeconomy/.

Reed, M.S. (2008). 'Stakeholder participation for environmental management: A literature review', *Biological Conservation*, vol. 141, no. 10, pp. 2417–2431.

Reed, M.S., Graves, A., Dandy, N., Posthumus, H., Hubacek, K., Morris, J., Prell, C., et al.. (2009). 'Who's in and why? A typology of stakeholder analysis methods for natural resource management', *Journal of Environmental Management*, vol. 90, no. 5, pp. 1933–1949.

Reed, M.S., Vella, S., Sidoli del Ceno, J., Neumann, R.K., de Vente, J., Challier, S., Frewer, L., et al. (2018) 'A theory of participation: What makes stakeholder engagement and public engagement in environmental management work?', *Restoration Ecology*, vol. 26, pp. S7–S17.

Republic of India, Ministry of Science and Technology (2014) 'National Biotechnology Development Strategy 2014'. Available at: http://dbtindia.nic.in/wp-content/uploads/2014/05/NBDS_2014.pdf

Republic of South Africa, Department of Science and Technology (2013) 'The Bio-Economy Strategy'. Available at: www.pub.ac.za/wp-content/uploads/2018/10/Bioeconomy-Strategy.pdf

Richards, C., Blackstock, K.L. and Carter, C.E. (2004) 'Practical approaches to participation', SERG Policy Brief No. 1. Aberdeen: Macauley Land Use Research Institute.

Ronzon, T., Santini, F. and M'Barek, R. (2015) 'The bioeconomy in the European Union in numbers. Facts and figures on biomass, turnover and employment'. Available at: https://ec.europa.eu/jrc/sites/jrcsh/files/JRC97789%20Factsheet_Bioeconomy_final.pdf

Rowe, G. and Frewer, L. (2000) 'Public participation methods: A framework for evaluation in science', *Technology and Human Values*, vol. 25, pp. 3–29.

Ruth, M. (2009) 'The nature of the beast and the beast in nature: Broadeing the perspective of technology', *Bulletin of Science, Technology & Society*, vol. 29, no. 5, pp. 374–382.

Sanz-Hernandez, A., Estaban, E. and Garrido, P. (2019) 'Transition to a bioeconomy: Perspectives from social sciences', *Journal of Cleaner Production*, vol. 224, pp. 107–119.

Schmid, O., Padel, S. and Levidow, L. (2012) 'The bio-economy concept and knowledge base in a public goods and farmer perspective', *Bio-based and Applied Economics*, vol. 1, no. 1, pp. 47–63.

Sirisaengtaksin, A. (2017) 'The update of bioeconomy roadmap.' Paper presented at AGBIO 2017. Available at: http://agbio2017.com/program/schedule/

Slunge, D., Drakenberg, O., Ekbom, A., Göthberg, M., Knaggård, Å. and Sahlin, U. (2017) *Stakeholder Interaction in Research Processes: A Guide for Researchers and Research Groups*. Gothenburg: University of Gothenburg.

Staddon, S.C., Nightingale, A. and Shrestha, S. (2015) 'Exploring participation in ecological monitoring in Nepal's community forests', *Environmental Conservation*, vol. 42, pp. 268–277.

Stadler, T. and Chauvet, J.M. (2017) 'New innovative ecosystems in France to develop the bioeconomy', *New Biotechnology*, vol. 40, pp. 113–118.

Tantiwatthanaphanich, T. and Zou, X. (2016) 'Empowering the local community via biomass utilization: A case study in Thailand', *International Review for Spatial Planning and Sustainable Development*, vol. 4, no. 2, pp. 30–45.

Tewdwr-Jones, M and Allmendinger, P. (1998) 'Deconstructing communicative rationality: A critique of Habermasian Collaborative Planning', *Environment and Planning A: Economy and Space*, vol. 30, no. 11, pp. 1975–1989.

Tippett, J., Handley, J.F. and Ravetz, J. (2007) 'Meeting the challenges of sustainable development: A conceptual appraisal of a new methodology for participatory ecological planning', *Progress in Planning*, vol. 67, pp. 9–98.

UNCED (1992). *Nations of the Earth Report*, vols 1–3. Geneva: United Nations.

van Tatenhove, J. and Leroy, P. (2003). 'Environment and participation in a context of political modernisation', *Environmental Values*, vol. 12, no, 2, pp. 155–174.

Vlaanderen.de (2014) 'Bioeconomy in Flanders. The vision and strategy of the Government of Flanders for a sustainable and competitive bioeconomy in 2030'. Available at: www.vlaanderen.be/publicaties/bioeconomy-in-flanders-the-vision-and-strategy-of-the-government-of-flanders-for-a-sustainable-and-competitive-bioeconomy-in-2030

Voulvoulis, N., Arpon, K.D. and Giakoumis, T. (2017) 'The EU Water Framework Directive: From great expectations to problems with implementation', *Science of the Total Environment*, vol. 575, pp. 358–366.

Wehn, U. and Evers, J. (2014) 'Citizen observatories of water: Social innovation via eParticipation', in *Proceedings of the 2nd ICT for Sustainability 2014 Conference*, Stockholm.

Woods, D. (2008) 'Stakeholder involvement and public participation: A critique of Water Framework Directive arrangements in the United Kingdom', *Water and Environment Journal*, vol. 22, no. 4, pp. 258–264.

Zeug, W., Bezama, A., Moesenfechtel, U., Jähkel, A. and Thrän, D. (2019) 'Stakeholders' interests and perceptions of bioeconomy monitoring using a Sustainable Development Goal framework', *Sustainability*, vol. 11, no. 6.

12 Conclusion and future directions

Udaya Sekhar Nagothu

Introduction

Historically, the cornucopian approach (the belief that resources are infinite) towards the environment prevailed among human civilizations. One good example is that of wood, which all along was an important resource for humans, used for construction, building boats, heating, cooking, agriculture and other purposes. In fact, this was one of the reasons for colonization, besides the demand for agricultural products, such as coffee, tea, spices and livestock feed in more recent years. The demand for timber and agricultural products resulted in large-scale deforestation, land degradation, soil erosion and loss of biodiversity in many parts of the world. On the contrary, we do come across several examples where communities lived, and continue to live in harmony with nature, until industrial agriculture replaced traditional management systems of natural resources. Local perceptions, beliefs and indigenous knowledge systems guided the societies to manage resources sustainably and avoid over-harvesting, for example, the way indigenous people of the Amazon lived in harmony with nature for centuries. Such examples are scattered and often limited to the local level. What we are witnessing today is the rapid depletion of resources and its implications reaching the global level in the form of climate change, famines and droughts.

The year 2019 witnessed some of the worst extreme weather events, unprecedented in the last century, particularly, the increase in average temperatures, which resulted in hotter days and thereby increased both the frequency and duration of heatwaves, as observed in Western and Central Europe (World Economic Forum, 2019a). A direct consequence of such extreme weather risks was on food production, as evident from the 15 per cent reduction in wheat yields in Germany in 2019. The year also recorded extreme heat, wildfires and rainfall in parts of South America, South Asia and Australasia. According to the World Economic Forum (World Economic Forum, 2019a), the global average temperatures were a little over 0.4 °C warmer in 2019 than the average between 1981–2010 (WMO, 2019). Any further increase can disrupt weather patterns across the globe. A slight shift in the monsoon patterns significantly affects agriculture, food security and the livelihoods of millions of farmers and rural people living in South and South-east Asia. More vulnerable are the regions and communities that are not prepared to deal with the impacts of extreme weather events that are likely to increase in intensity and frequency. These regions lack: (1) the necessary infrastructure; (2) the measures to combat risks to life and property; and (3) a proper strategy and policy.

The recent report published by the Intergovernmental Science-Policy Platform on Biodiversity and Ecosystem Services (IPBES, 2019) portrays an ominous picture of the deteriorating biodiversity and ecosystems on the Earth (United Nations, 2019). The report further warns that the consequences will be severe if the species and ecosystems are not protected, as it erodes the fundaments of the human economies, livelihoods and food security. The figures in the report portray rather a dismal picture of the global biodiversity scenario. The only option left to conserve ecosystems is through *a transformative change* and a system-wide reorganization of the way we manage our natural resources at the national and global levels. This may be a daunting task for scientists and citizens at large with the current state of political shortsightedness, lack of acknowledgement of science and the seriousness of the problems we are facing today.

At the same time, we need to bring millions of people out of poverty and feed the growing population and provide them with clean water and sanitation with minimum impacts on the environment. It must be done during a period of growing tension between development imperatives and the limits of the planet's resources. Heimann (2018) argues that a sustainable bioeconomy pathway assumes strong sustainability measures that reveal the extensive potential of the bioeconomy approach to support the achievement of the SDGs and address some of the grand global challenges. One of the major drivers of bioeconomy is to provide food and nutrition security with minimum impacts on the ecosystem.

The September 2019 fires in the Amazon and the political debate around it have drawn the attention of disparate community of scientists, activists and business groups who believe that technology or science can save the Amazon rainforests and promote sustainable development (*Financial Times*, 2019). Large-scale deforestation and land degradation can further aggravate global warming, crop failures and food scarcity. Where do we draw a line, and who should take the responsibility for managing the natural resources sustainably? The sustainability debate has been ongoing for nearly four decades now, and still we are no clearer about the policy and actions that are necessary for sustainable development. The changes we are advocating may not be adequate, as we need a transformative change to sustain a healthy ecosystem and at the same time ensure human well-being and sustainable economic growth. What exactly we mean by transformative change? According to the WRI (2017), a transformative change would require a shift in *what* we do, and not just *how* we do what we do. To begin with, a transformation of our lifestyles, the way we manage natural resources, what and how we produce and consume, at the individual, collective, national and global levels.

There are diverse and fragmented views within the scientific community and policy-makers on how we should use and manage natural resources in the future. According to the WEF report (2019b), the current production and consumption of resources are linked to economic rather than population growth, and technology, demand and supply, and market factors will dominate and influence the future pathways of production and consumption. Interlinked within this resource production and consumption cycles are the political, social and cultural factors that will play an important role in making future decisions linked to the use and management of resources. Unless sustainability becomes a fundamental criterion in policy and practice, protecting the Earth's resources for future generations will not be possible.

Sustainable bioeconomy, the SDGs and different perspectives

In simple words, business as usual will not suffice, looking at the rate at which exploitation of resources, land degradation, pollution of the environment, notably, the rivers, air and oceans are happening world over. The emergence of new paradigms is always positive, provided there are no vested interests and they can protect the vulnerable communities and sensitive ecosystems. This book, therefore, attempts to show that 'sustainable bioeconomy development' can be one such paradigm that has the potential to change 'what' and the 'how' we produce and consume bio-based resources while reducing the negative impacts on the environment. As discussed in Chapter 1, bioeconomy does not offer solutions to all the global problems and it comes with risks that need to be carefully addressed. According to some studies, it is another paradigm that has been introduced to meet the interests of the large corporates and industrialized nations (TransNational Institute, 2015). It is important to understand the relation between bioeconomy and sustainable development, and other similar concepts such as the green economy (Birner, 2017).

In recent years, an intersection between bioeconomy and the circular economy is gaining importance (Carus and Dammer, 2018). There is a need to examine its opportunities and limitations and develop more evidence-based approaches. The EU is one of the first regions to prepare a new action plan on the 'circular bioeconomy' approach that will enable it to launch several pilot actions across the EU (EC, 2018). The costs involved will increase, as circular (bio)economies seek to expand. These involve not just 'circulation costs' – in effect, recycling and reproduction costs – but also vastly increased transaction costs, which countries should be willing to pay. Technological advances may reduce both types of costs, but the question is, how quickly? The sustainability of bioeconomy technologies and innovations and whether they embody a footstep in the right direction towards achieving a circular bioeconomy should be examined before they are approved (Issa, 2019). Countries such as Finland have initiated an ambitious policy to promote circular bioeconomy connected mainly to the forestry sector. However, innovative, agile, flexible and encouraging organizational culture is a crucial resource for putting the policy into practice (Nahya, 2019).

It is an ambitious target to achieve the UN SDGs by 2030. With the inconsistent policies, fewer investments and lack of commitment, it may not be easy to achieve the targets within the next 10 years at the global level. In fact, sustainability is emphasized in the EU's definition of bioeconomy and followed by the member countries. EU also considers bioeconomy as a potential pathway to achieve the SDGs. However, there are several studies that do not agree with the sustainability dimension of bioeconomy policies and there are compound tensions between the two dimensions (Global Bieoconomy Summit, 2015; Juerges and Hansjürgens, 2016). Some critics believe the current concept of bioeconomy does not necessarily meet the criteria of the SDGs (German Bioeconomy Council, 2015). One must not ignore the risk of conflicting interests within the SDGs, for example, an attempt to address SDG 2 can have a negative impact on SDG 15, if sustainable and smart agriculture systems with minimum impact on the terrestrial ecosystems are not implemented. The Green Revolution was in fact criticized for its negative impacts on land, water and biodiversity, despite its positive contribution to address food security and hunger eradication (Nagothu, 2015). On the other hand, SDGs, the associated indicators and monitoring systems, complementarily, can be used as sustainability benchmarks to measure the success of bioeconomy development (Heimann, 2018). It is how the governments and stakeholders must weigh up the different options, in the pursuit of achieving several SDGs. According to Dietz et al. (2018), sustainable use of bioresources can help to achieve the SDGs that in turn can improve the social, economic and ecological conditions.

There is also an increasing commitment by several governments, as observed in Germany, where the potential of sustainable bioeconomy in the achievement of SDGs is acknowledged and explicitly stated in its Bioeconomy Evaluation Report, 2017. The SDGs' framework thus can provide a legitimate platform to promote bioeconomy approach. However, some critics do not agree that SDGs and the targets set by the UN can provide any legitimate framework for cooperation (Easterly, 2015).

Despite these contradictions, sustainable bioeconomy development has gained importance over the last decade among scientists and policy-makers as one of the promising alternative paradigms to address the grand global challenges (Bugge et al., 2016). It is necessary to improve the clarity in the concept and what it implies in practice. Three broad perspectives or views on bioeconomy emerge from the literature review done by Bugge et al.

(ibid.), including, first, a bio-technology view, second, a bio-resource view emphasizing processing of bio-based materials and value chain analysis, and, third, a bio-ecology view emphasizing sustainability and maintenance of ecosystem services. The majority of the bioeconomy proponents currently appear to be revolving around the bio-resource view where economic growth is given importance. Here, the emphasis is on the value chain enhancement to promote efficient use of bio-resources to develop new products, minimize waste in the production process and increase the circularity of resource use. In addition, stakeholder engagement, promoting innovation and research are also seen as important elements in the bio-resource view. Any future bioeconomy strategy should ensure it checks: non-food uses of arable land, use of crop land to produce fodder and feed for meat, milk and egg production and, finally, should not allow the conversion of virgin forests or sensitive ecosystems to agricultural land (Issa, 2019).

In general, the main motivation for sustainable bioeconomy development, as observed in most of the prevailing strategies, is to achieve one or more of the following objectives: (1) the need to address the societal or the grand global challenges, including climate change; (2) ensure food and nutrition security; (3) increase circularity and efficiency in resource use and production, and reduce GHGs; and (4) promote economic growth and sustainable development (Meyer, 2017). What is missing and most crucial is the emphasis on healthy ecosystems and sustaining the ecosystem services, on which the long-term success of sustainable bioeconomy will depend. Most innovative bioeconomy technologies are currently at different stages of development. As we progress, there is a need for standardization of the technologies, sustainability criteria and their measurement, economic and social feasibility and affordability by communities.

At the global level, the G8 countries have been addressing the challenges of climate change, global food security and sustainable development and thus touching on the important goals of bioeconomy. However, the focus in the field of bioeconomy is still at a national level, except in the EU, where a common bioeconomy strategy has been developed. Groups such as the G8 or G20 have the potential to open doors to link the regional bioeconomies and develop a common global bioeconomy framework where countries can collectively contribute to the implementation of the 2030 Agenda for Sustainable Development and its Sustainable Development Goals (SDGs) (SEI, 2018). However, the global bioeconomy agenda should benefit the larger global community interests and ensure transparency (TransNational Institute, 2015). Emerging economies such as China, India, Brazil, South Africa, Thailand and Malaysia have great potential, if they can develop innovative technologies and policy that can increase the value and efficiency of biomass resource use, particularly the bio-residues or waste and promote circularity in the process.

This final chapter in the book summarizes the main messages from the various chapters. In the introduction, a critical view of the sustainable bioeconomy concept and the relevance of sustainability is provided. This is followed by short abstracts of each chapter highlighting the important concepts and findings from different case studies. In the later part of the chapter, conditions to overcome constraints for implementing bioeconomy and main lessons learnt from selected chapters are discussed. Finally, the chapter provides some policy recommendations and concluding remarks.

Brief summaries of chapters

Chapter 1 offers a comprehensive literature review on sustainable bioeconomy development, with a focus on the definition, concepts, potential and risks involved. Differences exist between the countries on how they view bioeconomy, with some putting emphasis on sustainability and ecosystem services, while others on the economic growth as the main goal. The literature review shows that bioeconomy and circular bioeconomy are rather new concepts, at times with their goals conflicting, and objectives opposing. However, bioeconomy brings the sustainable development discussions back onto the policy agenda both at the national and international levels. Lack of a common bioeconomy agenda and understanding across the globe will be one of the main constraints to achieve SDGs. Experiences so far are limited, and hence future bioeconomy development must be based on strictly responsible, accountable and sustainable use of natural resources, but at the same ensure human well-being and economic growth.

Chapter 2 focused on ocean-land interactions and implications for bioeconomy that offers unique opportunities to feed the growing human population. Oceans can provide a circular bioeconomy by utilizing increased CO_2, and dissolved nutrients (P, N, Fe and other elements) in the water, leached from land-based activities. Estimates show that CO_2 capture by seaweed cultivation alone can range from 3,000 to 1,500 tons per square kilometre. Ocean photosynthetic production provides more food and energy for human consumption without external inputs. This can contribute to sustainable development – provide food security and recovery of degraded ecosystems, thus contributing directly to SDG 2 (reducing hunger) and SDG 14 (protecting life below water). Nevertheless, increasing food production from the oceans has its associated risks if proper conditions are not created. Hence, proper coastal land use management is important as it continuously affects the nutrient flows, which in turn can lead to stronger changes in carbonate chemistry and ocean acidification. Genuine and stable partnerships, therefore, are necessary to share responsibility for environmental stewardship and manage marine and coastal ecosystems sustainably. The chapter suggests the need for financial incentives to encourage research and innovations, support farmers' associations and establish common platforms to share data and knowledge on oceans for better management.

Chapter 3 emphasized the supply of nutrients through complex biogeochemical cycling that underpins all ecosystem services, which, in turn, provides the basis for bioeconomy. Iron, one of the essential elements for life, is usually in low concentrations in oceans compared with lands. Thus, iron supply from lands to oceans is a crucial process to foster phytoplankton growth in oceans. Recent studies show that the Amur River is supplying a huge amount of iron to the Sea of Okhotsk, and thus plays an important role in supporting phytoplankton growth of the adjacent Oyashio Region. Hence, the Amur-Okhotsk ecosystem was named the 'Giant-Fish Breeding Forest' (GFBF), inspired by Japanese traditional word '*Uotsukirin*'. Since large uncertainties remain in the system, not only continuous research, but also careful management of the system are required. The Amur-Okhotsk Consortium, established in 2009, must enhance communication and support to address conflicts among stakeholders and help in better management of the ecosystem.

In *Chapter 4*, the importance of sustainability of local resource use and a comparison of patterns of production and consumption in selected prefectures in Japan were discussed. The results from principal component analysis (PCA) conducted in the study showed a strong association between population density, gross domestic product (GDP) and resource use in general. PC1 was negatively correlated to population density but showed a positive relationship with per capita natural capital and water use, whereas PC2 was negatively related to per capita GDP, ecological footprint, per capita carbon emissions and electricity demand, thereby, suggesting a gradient of economy and energy use. PC3 was negatively related to agricultural production and self-sustainability of food. Per capita proportion of renewable electricity, per capita carbon sequestration, self-sufficiency of water, food and cultural services (such as field school destinations of junior high schools) were high in prefectures with less dense populations. Overall, the study revealed that rural areas have higher resource self-sustainability, though economic power and population are concentrated in urban areas. The study suggested that rural areas must be compensated for the cost of production of renewable resources and ecosystem services they generate. The main message is that the 'local ecological footprint' based on self-sufficiency is an important index for local sustainability evaluation which in turn contributes to measuring the national or 'global footprint'.

Chapter 5 highlighted the challenges in the agriculture sector in Africa and showed that the current systems are not productive enough but are rather linear, dependent on fossil fuels, and, on the contrary, depleting natural resources. The chapter reviewed the potential of sustainable intensification of agriculture with emphasis on diversified cropping systems and value chain enhancement as an option to promote the bio-based economy in rural regions of Africa. The chapter used data and experiences from an ongoing research programme in Malawi (www.innovafrica.eu), where maize-legume cropping systems were introduced for adoption by smallholders. There is a great potential to apply the 3Rs principle of bio-based economy (i.e. reduce, reuse and recycle) in the farming systems at production, post-production, marketing and processing stages of the value chain. To sum up, the sustainable intensification approach inclusive of value chain development, appears to be a promising option for smallholders in Sub-Saharan Africa, which can improve productivity, farmers' income, improve gender mainstreaming and at the same time reduce environmental impacts.

Chapter 6 summarized the recent work by the FAO and its national partners on sustainable intensification of rice crop production in South-East Asia. The chapter presented new research findings on farmers' adoption of good practices resulting in an increase in productivity and profitability of smallholder farming systems that can be directly attributed to Save and Grow (S&G) farmer training interventions in rice-based landscapes in the Lao PDR, Vietnam and elsewhere in the Greater Mekong subregion. The case for increased policy support – and public-private sector investments – for sustainable intensification of crop production and education programmes for smallholder farmers was outlined in the chapter. The aim was to develop more *inclusive* 'Green Rice Landscapes' while contributing to the achievement of multiple Sustainable Development Goals (SDGs). Such investments are also vital for the development of robust and resilient rural economies that are based on key circular bioeconomy concepts, principles and good practices, as outlined earlier in this book. The chapter also aligns well and provides real-world examples of successful transformation towards greater sustainability in the agriculture sector as called for in recent landmark publications by the United Nations.

In *Chapter 7*, protected cultivation (PC) and its potential for smallholders in Sub-Saharan Africa (SSA) were discussed. PC is defined as the protection of crops against adverse climatic and biological factors using physical means or structures. As natural resources become scarcer, farm inputs become more expensive and consumers more attentive to the sanitary and nutritious aspects of food produced, PC has become one promising yet improvable way to address the challenges. As the global demand for high value crops such as vegetables increases, PC provides an opportunity for farmers to increase their income, improve their working conditions and 'de-risk' their production, thereby attracting youth in a crucial but ageing sector. In SSA, farmers constantly face the threat of losing their crops due to pests, diseases or harsh climatic conditions, with no insurance to cover the loss. A case study, comparing PC of tomatoes in Benin and Tanzania, showed that PC was profitable despite high initial investments. However, it varies depending on the market for high quality produce. Favourable policies, farmers' training and agronomic research are prerequisites to the adaptation and development of PC technologies and techniques in SSA. PC has the potential to enhance bioeconomy in the region and at the same time achieve some of the SDGs.

The case of Göksu, Anatolia, in Turkey and the way humans traditionally have managed natural resources and how land use was shaped over the centuries based on the potential and human needs were presented in Chapter 8. The territory was never rich in fossil fuels. Settlers in the region therefore relied on wood to meet their fuel demand. This gradually led to severe erosion problems, particularly, along the Mediterranean coast. Although limitations existed, the locals developed traditional management practices for securing their food and fibre production by re-shaping the environment, which is termed an 'anthroscape'. The anthroscape approach adopted in the chapter reviewed the land use and the potential for bioeconomy development, when resources are sustainably used. The chapter also explores to what extent land resources in Anatolia can support the new renewable energy targets and other biomass needs in Turkey. It suggested measures to improve the availability of biofuel resources and develop other sectors that are now dependent on bio-resources in the region. The chapter analyzed the net primary production, the total organic carbon and humic acid content of different land use types in Anatolia, which give an indication of the ecological sustainability of the anthroscape. Information about production potential and the sustainability of different land use types in the region is necessary for better land use planning.

Chapter 9 presented the case of sustainable woods management in Japan, and the multiple benefits it can generate for the rural communities. The Nishiawakura village, located in Okayama Prefecture, is a typical example in Japan where plantation forests are matured and under-utilized. The village community decided to revitalize the area through sustainable utilization of natural resources. In 2008, the village decided to adopt the 'Centrum Woods Vision' initiative, to develop a new value chain of wood resources and increase the production and use of renewable energy. Young people within and outside the village supported the initiative and contributed to this development as entrepreneurs. Nishiawakura's pursuit of sustainability through multi-faceted activities based on natural resources is at the forefront of the upcoming bioeconomy-oriented society. The initiative simultaneously pursues several of the SDGs, including Goal 5 (gender equality), Goal 7 (affordable and clean energy), Goal 9 (industry, innovation and infrastructure), Goal 10 (reduced inequality), Goal 12 (responsible consumption and production), Goal 13 (climate action), Goal 14 (life below water) and Goal 15 (life on land).

Chapter 10 analyzed the main challenges and opportunities to promote sustainable biogas technology adoption by smallholders through integrated food and energy systems (IFES), using a case study from Malonga village in the Limpopo Province of South Africa. Biogas has become attractive in recent years because of its multiple benefits and contributions it can make to the SDGs. However, in Africa, its adoption remains low, due to several constraints, including: (1) lack of access to feedstocks and water scarcity; (2) high initial/upfront cost of installation and lack of investments; (3) lack of skilled labour for installation, operation and maintenance; (4) limited training facilities; (5) inadequate policy support and extension services; and (6) slow behavioural and social acceptance. Based on the baseline information collected, an integrated framework condition that can improve the adoption of smallholder biogas technology through IFES was suggested. IFES will only succeed in delivering, if the necessary framework conditions such as adequate feedstock and water, training, policy support, stakeholder collaboration, credit and insurance and support services are provided. The implementation of the necessary framework conditions for biogas technology should be underpinned by conducting an integrated research study using type 2 IFES in the context of smallholder farmers in Africa.

Chapter 11 emphasized the need for active stakeholder engagement right from strategy development to planning and implementation to realize the benefits of sustainable bioeconomy development. In general, this varies between regions and countries. In the EU, it is considered important to engage stakeholders at all stages, whereas in developing countries engaging stakeholders has not received much attention so far while launching new strategies. Stakeholders, including the private sector, research institutions, farmers organizations, government and non-governmental organizations, all have important roles to play. The chapter focused on why, how and what type of stakeholders should be engaged, and the relevant benefits and challenges that can be derived. It discussed experiences from the EU and other regions where stakeholder engagement (formal and informal) and participative governance have led to or are necessary for a successful and sustainable bioeconomy development.

Bioeconomy and sustainable development: the key messages

In this section, some of the important messages and indicators for sustainable development that emerged from the analyses in different chapters will be highlighted, and their relevance to bioeconomy discussed.

The importance and potential of oceans as a source of food for the future were highlighted in Chapter 2 and Chapter 3 in the book. We can produce more food from the oceans, provided we do so in a sustainable manner. In the process, it must be ensured that the ecosystem services are maintained while we exploit the resources for biomass production. In these two chapters, the importance of ecosystem services in the form of biogeochemical cycling, and the dynamics of nitrogen, phosphorus, iron and other minerals transported through these processes from land to ocean were discussed. In the case of the Amur River and the Sea of Okhotsk and the Oyashio region, which constitute one large ecosystem, iron transportation was observed to be crucial to the maintenance of ecosystem services and ocean biomass production, including fish and aquatic plants. In Chapter 3, the concept of '*Uotsukirin*', meaning 'fish-breeding forest' was highlighted, which explains the functionality of the land-ocean system interactions. People in the Edo period in Japan empirically found that fish gathered under the shoreline where trees were growing. This traditional knowledge was later formalized to express the specific conservation zone in the modern forest legislation of Japan. Thus, tracking the relationships between traditional knowledge and modern economic activities might give a useful insight into the future bioeconomy planning and development. In the process, it might motivate the communities to manage their resources sustainably, avoid over-harvesting, and consider environmental protection as one of the priorities.

Sustainability should be monitored and should become an integral part of planning and management to check the over-use of resources. Chapter 4 suggested the importance of local sustainability and the *local ecological footprint*. The outputs from the analyses showed that we must not only have discussions on resources merely for commercial use, but also combine them with resources that are important for sustainable ecological and social well-being. One of the recommendations from the chapter was that ecosystem services can provide important indicators and tell us about the health of the ecosystem we depend upon and how we need to change or adjust the production and consumption patterns to follow a sustainable development path. Thus, measurement of local ecological footprint is highly relevant and a must for the regions that are pursuing bioeconomy development in the future. Future strategies developed must be embedded in a sustainable framework right from the beginning.

Local sustainability has a direct impact on urban areas and vice versa, as, these days, cities receive much of their resources (fresh water, forest and agricultural produce, tourism) from outside, either from the rural areas within the same country or from other countries. However, the environmental costs to produce or manage these resources elsewhere are not paid by cities. Therefore, Chapter 4 suggested that mechanisms should be in place to ensure that the management costs of ecological resources are also covered by the people who consume these resources. In this context, the relevance of instruments such as the *payments for ecosystem services* (PES) was highlighted. PES was suggested as one of the instruments to support sustainable development, both in the cities and rural areas. Simple and easy-to-operate PES examples should be recognized, based on ongoing experiences.

Sustainable agriculture intensification systems were demonstrated as a viable option to increase productivity and farm income with reduced environmental impacts in Chapter 5, Chapter 6 and Chapter 7. Adoption of sustainable agriculture systems is expected to reduce the chemical inputs, reuse the resources produced and recycle the resources or bio-residues that can boost circular bioeconomy and improve productivity and efficiency in the agriculture sector. This is highly relevant for most developing countries and emerging economies including countries in Africa,

South and South-East Asia and South America where resource use efficiency and productivity are well below the potential levels. These regions need to increase food production to feed their growing populations while minimizing environmental impacts. One of main recommendations from Chapter 5 was to *promote diversified cropping systems*, for example, maize or millets with legumes that can benefit both human nutrition as well as improve soil fertility, as compared to monocultures. For this to succeed, the necessary institutional conditions and an enabling policy framework should be provided together with suitable technologies for smallholders, for example, affordable seed supply systems, targeted extension services, storage and market facilities and fair and stable farm gate prices.

Chapter 6 demonstrated the positive impacts of the Regional Rice Initiative – *Save and Grow* farmer education programmes implemented through National Integrated Pest Management (IPM) Programmes and upscaled by Farmers Field Schools. Farmers learn and test good practices that are more productive, sustainable and efficient in resource use while promoting better *agro-ecology-based intensification and diversification of rice-based farming systems*. The integrated system approaches included: rice-fish, rice-livestock and rice-vegetables/potato diversified production systems, IPM, agro-forestry and promotion of climate-smart farming systems. The overarching Save and Grow paradigm is aimed at moving agricultural development away from a focus on single crop commodities and focal areas – e.g. improved seed, pest control and water management – to more holistic agroecology-based solutions that promote diversification (FAO, 2014, 2015). This paradigm is aligned with the circular bioeconomy concept, principles and good practices. Thus, crop diversification in fact helps in the efficient use of resources, minimizes chemical inputs and provides multiple income sources for the farmers.

Another useful concept namely, *anthroscapes* was discussed in Chapter 8. Attempts were made to quantify the carbon dynamics of the anthroscapes in the study and a new approach to develop regions based on carbon accumulation was suggested. Land-use assessment was done using indicators such as the *net primary production* (NPP), *total organic carbon* (TOC) and *nitrogen* (N) in the soil. These indicators and their measurement methods are important to analyze the impact of different management patterns on land quality (Zhang et al., 2013). Such concepts and indicators will be important for regions that want to pursue bioeconomy development in the future. These indicators can measure the impacts we will create on land and soil characteristics as we change land use to meet our economic needs. The chapter also suggested using *soil humic acid* content as an indicator to assess the biomass-supporting ability of soils, as soils with high humic acid are likely to support high NPP that may in turn increase the bioeconomic domain's economic value (Olivares et al., 2017). The attempts to increase humic acid in the soils will also mean incorporating more organic matter into the soils, such as crop residues, that can increase water-holding capacity, aeration and soil biodiversity. The biomass-production-based crop/soil management practices in these domains will enhance CO_2 *sequestration*, essential for sustainable bioeconomy development.

The '*Centrum Woods Vision*' initiative adopted in the Nishiawakura village of Okayama prefecture of Japan, as discussed in Chapter 9, offers important experiences that can be useful to similar situations elsewhere in Japan and the rural areas of other countries. It demonstrated how young people from within and outside the village were given an opportunity and provided with the necessary conditions to *improve the wood value chain* and promote the use of renewable energy, utilizing natural resources around the village. In the case of Nishiawakura, aspiring entrepreneurs were assisted by the 'Community revival cooperation corps' system, a scheme implemented since 2009, by the Ministry of Internal Affairs and Communications, Japan, to support disadvantaged regions (Ministry of Internal Affairs and Communications, 2019). Thus, the scheme is proving to be effective in promoting bioeconomy-related activities, encouraging people to move to depopulated rural areas, creating job opportunities and giving them time to adapt to the different lifestyles.

This is a living example of how national policy support can attract youth to live in rural areas and can slow the migration to the cities. As shown in the Nishiawakura village, sustainable bioeconomy development with the necessary conditions in place can offer a promising pathway to achieve rural economic growth. The case has also demonstrated how genuine stakeholder cooperation at the local level has resulted in a successful outcome for the village community. As youth opt to move to rural areas, it is important to motivate them by providing educational facilities for their children, affordable housing and renewable energy options for domestic use and faster transport facilities.

Indicators for monitoring and measurement of sustainability

As already indicated in Chapter 1 of this book, there are standard indicators that can be used to measure the contribution of new paradigms such as bioeconomy to the SDGs. Indicators to measure sustainability at a macro-scale have been in existence since the 1980s or early 1990s (Barbier and Markandya, 1990). However, no one set of indicators can be used across all regions or countries with different standards, ecological and socio-economic settings. Hence, there needs to be flexibility and region-specific standards in monitoring the contribution of bioeconomy. However, the measurement methods could follow standard procedures.

It is necessary to have the right infrastructure and procedures to measure and monitor sustainability criteria of any new bio-based initiative that impacts natural resources. New digital or ICT-based tools could help to collect data and support the evaluation of sustainability.

In the Nordic region, the measurement of bioeconomy impacts is based on criteria including: the sustainable use of resources; technology and business model innovation, environmental and societal benefits. Any new bioeconomy initiative and its impacts in the region are measured based on these criteria. Such monitoring keeps a check on the ecological, economic and social sustainability of the bioeconomy programmes. Measurement and monitoring are mostly done by independent bodies, government agencies or scientific institutes on a regular basis and reports are issued. Sustainability ecological indicators, such as low ecological footprint, total organic carbon, and nitrogen, discussed in the book, need to be monitored on a regular basis to measure the health of the ecosystem and change the management accordingly. Similarly, economic and social indicators including household income, livelihood and social well-being are important for sustainable bioeconomy development. Table 12.1 shows a list of selected case studies discussed in the book and their potential to contribute to the sustainability (in the different columns). These cases cover different sectors including fisheries, agriculture, forestry and bioenergy.

Table 12.1 Some relevant concepts discussed in the book and their contribution to bioeconomy criteria/sustainability dimensions

	Environmental sustainability	*Economic sustainability*	*Social sustainability*
	Productivity/ ecosystem services	*Providing jobs and income*	*Increased capacity and welfare*
Giant Fish Breeding Forest (bioresources from water)	+++	++	+
Sustainable agriculture system/ value chain enhancement	++	++	++
Save and Grow Sustainable Intensification of Agricultural Production policy and associated good practices	++	++	++
Protected vegetable cultivation	+	++	+
Anthroscape approach for land use	+	++	+
Centrum Wood Vision (new value chain of wood and renewable energy)	+++	++	+
Sustainable bioenergy/biogas initiatives	++	+	+

Notes

+ denotes some possibility; ++: high possibility, +++ very high possibility.

Wesseler and von Braun (2017) reported that measuring bioeconomy impacts can be addressed through systematic data collection and analysis, developing baseline indicators and economic tools for measurement. The methods for evaluating sustainability should be robust, not data-hungry and easy to measure without the need for complicated and expensive equipment. Many countries do not give access to data, and, on the contrary, scientists are required to purchase data from government and other agencies, which is a major constraint. Therefore, governments should make data access and sharing easier for scientists whenever needed. At the same time, technical capacity to monitor and measure sustainability indicators should be strengthened through regular hands-on training, funds to organize training courses and cross-sectoral cooperation.

Possible measures to overcome constraints/or risks associated with bioeconomy

According to Kleinschmit et al. (2017), bioeconomy cannot be considered 'self-evidently sustainable'. It comes with several risks, including technological, environmental, economic and policy-related ones (European Parliament, 2017). Bioeconomy-based sectors pose challenges due to significant shortages of biomass stocks, limited technology, and unstable markets for the newly developed products (Camilla et al., 2017). Lack of appropriate technology and cost factors in developing countries often force compromises on the quality and principles of sustainability that can lead to negative environmental impacts. Therefore, incentives must be in place to support marketing of products that are produced sustainably. Initiatives such as the *green labelling* of products, increasing awareness in the consumer market, and monetary mechanisms to compensate some of the production costs will help. But we need much more than that in today's context, where globalization and trade impacts cross international borders. It is thus necessary to increase transparency all along the value chain, including the impacts on health of ecosystems and costs for local communities.

In bioeconomy, both 'enabling' policies that provide incentives and measures to promote bioeconomy, and 'restrictive' policies to control or regulate the use of bio-resources should be in place (Dietz et al., 2018). Both are necessary to ensure sustainability in practice, but at the same time they are politically challenging (El-Chichakli et al., 2016). Despite the policies, stakeholders may not put them in practice due to overhead costs and short-sighted profit motives. Hence, a policy framework that is implementable, targets that are achievable and environmental standards that are necessary will be useful for a sustainable bioeconomy approach. The policy framework should be able to take care of the environmental and social externalities (Zilberman et al., 2013).

Lack of international consensus on which products and activities are within the bioeconomy is another constraint to be addressed (FAO, 2018).

At the same time, it is not easy to systematically monitor bio-based production due to its scattered nature and often being driven by small-scale producers, especially in developing countries where they constitute the majority. Consequently, its contribution to the national Gross Domestic Product (GDP) can be underestimated and may not be attractive for politicians to give their support. With the *digitalization* options available these days, simple *mobile-based apps* can connect small-scale producers to report and provide timely data on production and sustainability indicators. These apps can also help to provide green labelling and monitor the ecological footprint. In a market where the consumers are becoming conscious about the ecological impacts of goods and services, online data about producers will help them to get connected and buy their produce directly and by-pass middlemen. There are several good examples of such producer-consumer networks in the peri-urban areas in the EU, Australia, and the USA, directly marketing vegetables, fruits, fish and other fresh agricultural produce.

The cross-sectoral nature of bioeconomy demands that all sectors must be engaged from the beginning. This must start with the establishment of an inter-ministerial body involving all relevant ministries in planning development and implementation of the bioeconomy strategy. As mentioned earlier in Chapter 1, countries such as Germany have already set up such inter-ministerial bodies to oversee bioeconomy activities. Another example is that of Japan, where the National Biomass Policy Council was set up with representatives from seven different ministries including, Economy and Trade, Forestry, Agriculture, Internal Affairs, etc. Opportunities for stakeholder and public engagement in the governance of the bioeconomy occurs at the regional level (Hasenheit et al., 2016). This may not be easy, given the competition for bio-feedstocks between users. Hence, adequate resources should be made available at the regional level or even local level to establish a stakeholder platform for engagement. We saw that in the case of Nishiawakura in Japan where local *stakeholder cooperation* has been one of the main reasons for success. This is also the main message from Chapter 11, which focused on *stakeholder cooperation and participative governance* as important aspects of bioeconomy. Thus, cooperation in bioeconomy research and development must be improved, across sectors, scientific agencies and governments in different countries. Joint programmes in research, and *open access knowledge platforms* for sharing available knowledge should be developed for better results.

Conditions necessary for sustainable bioeconomy development

It is always good to have a wish list of necessary conditions that must be in place for ensuring success. But the questions are: *what is possible?*, and if so, *who can provide them?* and *who is willing to take responsibility?* Some of these questions have already been pointed out in the previous section and in several of the book chapters.

First and foremost, scientists would want *innovation and research* to be an integral part of sustainable bioeconomy development. This would mean adequate funding, time and capacity for research, cooperation among research disciplines, easy data access, and sharing the results, knowledge or technologies developed. Scientific findings should at the same time support policy and practice. Adamowicz (2017) emphasized the need for increased research and continuity of research programmes, knowledge networks, and systems that can ensure technology and innovation uptake. We need standard methods for the assessment of bioeconomy potential, which should be developed for different situations, for example, on local carbon and ecological footprint assessment. Apart from research, standardization and certification are also necessary for energy, material and fuels.

The H2020 research programme supported by the EU is a good example of how research and innovation closely follow policy and strategies developed (EC, 2013), in this case, the EU Bioeconomy Strategy (EC, 2012). Similarly, in the Nordic region, funding is exclusively provided by the respective governments for research and innovation in bioeconomy-related topics. Whereas, in emerging economies such as Brazil, India, China, South Africa and Russia, no such specific targeted funding for bioeconomy research is yet available, although scattered funding instruments exist. In some of the countries, there is a thrust on funding biotechnology, targeted towards industrial production.

It is necessary to have proper *infrastructure and investments* to support sustainable bioeconomy initiatives. This could involve developing physical facilities for research, processing, storage, transport and marketing. This was one of the main recommendations in Chapters 5 and 6, where value chain enhancement at all stages was shown to be necessary in Africa to promote bioeconomy in the agriculture sector. The other prerequisite for bioeconomy is to create *functional markets* for the new products developed. In the EU, facilities are being set up to promote *bioeconomy hubs or clusters* from the ocean and land-based resources, especially in marginal areas where land is available and in abandoned public infrastructure in the rural and peri-urban areas. At the same time there are ongoing initiatives for *training and capacity-building* to develop adequate technical resources, including courses in the universities and short vocational training programmes. The agroecological concepts and associated good practices that farmers learn about in *Farmers Field Schools* also fit well into this category of capacity-building and learning needs for development of circular bioeconomy. The European Commission emphasized three elements when it launched the Bioeconomy Strategy in 2012 including investments in research, innovation and *skills development*, strengthening political coordination, and developing markets and competitiveness for sustainably developed bioeconomy products (EC, 2012). Lessons from the

EU during the last eight years can be useful for other countries trying to develop their bioeconomy strategies.

The proactive bioeconomy policy and government initiatives in the EU have already generated 22 million new jobs. Bioeconomy can open doors and create new job opportunities for *youth and women*, particularly in the emerging economies, where unemployment is a serious issue. Stable public-private partnerships between public and private agencies are becoming important to promote bio-based economy initiatives. It may not be easy in all situations to establish partnerships due to conflicting interests in the resources. Since there are several actors who own, govern, produce and consume the resources, cooperation between the actors or stakeholders, as emphasized in Chapter 11, is nevertheless becoming a necessity. This helps in creating ownership, transparency and responsibility towards promoting a sustainable bioeconomy. There is a need to increase awareness and social acceptance to implement bioeconomy in regions where it is new.

The majority view is that bioeconomy will have positive sustainability effects if suitable conditions are provided and appropriate measures are taken (Meyer, 2017). Sustainability should be seen as the goal and the guiding principle of bioeconomy development. However, to achieve sustainability, various technical, economic, and social prerequisites are necessary that the bioeconomy itself cannot create (Pfau et al., 2014). According to the German Bioeconomy Council (2017), a combination of both top-down and bottom-up initiatives, in the form of policy, technology and investment support would be required for successful bioeconomy development.

Conclusion

Sustainable bioeconomy development requires a fundamental change in the structure and functioning of the agencies responsible for implementation in the respective countries. In this context, cooperation between the research community, civil society, industry and policy-makers is very relevant. It requires a transformative change of the way we produce and consume resources, at the individual and collective levels. The latter refers to collective actions leading to cooperation at sectoral and regional levels and across countries. Given the current political scenario at the global level, economic crisis, and trade wars, collective global action will become even harder. This gives large room for initiatives based on individual consciousness to respond to the need for sustainable use of the Earth's resources. We see more and more of such movements, particularly, among youth and consumer groups to act responsibly.

There are always risks associated with any new paradigm, hence, it is necessary to ensure transparency in the process, consider the interests of the most vulnerable groups and genuine stakeholder management from the beginning. Several case studies in the book prove that bioeconomy can help to achieve several of the SDGs. This demands a global consensus and a common framework for sustainable bioeconomy development, which could make the global commitment to address SDGs even stronger. Sustainability indicators, including ecological (local ecological footprint, total organic carbon, soil nitrogen, biogeochemical processes and other ecosystem services), economic and social sustainability indices are important to measure and monitor regularly to ensure that we are on a sustainable development path. A healthy ecosystem is the only basis to ensure sustainable bioeconomy development, in which human well-being and economic growth can be achieved in the long run.

References

Adamowicz, M. (2017) 'Bioeconomy as a concept of development strategies in the European Union', *Journal of International Business Research and Marketing*, vol. 2, no. 4, pp. 7–12.

Barbier, B.E. and Markandya, A (1990) 'The conditions for achieving environmentally sustainable development', *European Economic Review*, vol. 34, nos. 2–3, pp. 659–669.

Birner, R. (2017) 'Bioeconomy concepts'. Available at: https://link.springer.com/chapter/10.1007/978-3-319-68152-8_3

Bugge, M.M., Hansen, T. and Klitkou, A. (2016) 'What is the bioeconomy? A review of the literature'. Available at: www.mdpi.com/2071-1050/8/7/691

Camilla, A.A., Gonzalez, R., Kelley, S. and Jameel, H. (2017) 'Risk management consideration in the bioeconomy'. Available at: www.fs.usda.gov/treesearch/pubs/55507

Carus, M. and Dammer, L. (2018) 'The circular bioeconomy, concepts, opportunities and limitations'. Available at: http://bio-based.eu/download/?did=120804&file=0 (accessed 13 March 2019).

Dietz, T., Börner, J., Förster, J.J. and von Braun, J. (2018) 'Governance of the bioeconomy: A global comparative study of bioeconomy strategies', *Sustainability*, vol. 10, p. 3190.

Easterly, W. (2015) 'The SDGs should stand for senseless, dreamy, garbled'. Available at: https://foreignpolicy.com/2015/09/28/the-sdgs-are-utopian-and-worthless-mdgs-development-rise-of-the-rest/

El-Chichakli, B., von Braun, J., Lang, C., Barben, D. and Philp, J. (2016) 'Policy: Five cornerstones of a global bioeconomy', *Nature*, vol. 535, no. 7611, pp. 221–223.

EC (European Commission) (2012) 'The Bioeconomy Strategy'. Available at: https://publications.europa.eu/en/publication-detail/-/publication/1f0d8515-8dc0-4435-ba53-9570e47dbd51

EC (European Commission) (2013) 'H2020 Bioeconomy'. Available at:, https://ec.europa.eu/programmes/horizon2020/en/h2020-section/bioeconomy

EC (European Commission) (2018) 'A new bioeconomy strategy for a sustainable Europe'. Available at: https://ec.europa.eu/commission/news/new-bioeconomy-strategy-sustainable-europe-2018-oct-11-0_en

European Parliament (2017) 'Bioeconomy: Challenges and opportunities'. Available at: www.europarl.europa.eu/thinktank/en/document.html?reference=EPRS_BRI(2017)595890

Financial Times (2019) 'Can science save the Amazon?', *Financial Times*, September 12, p. 7.

FAO (Food and Agriculture Organization) (2014) 'The multiple goods and services of Asian rice production systems'. Available at: www.fao.org/3/a-i3878e.pdf

FAO (Food and Agriculture Organization) (2015) 'Stories from the field: Vietnam, Save and Grow'. Available at: www.fao.org/climate-smart-agriculture-sourcebook/enabling-frameworks/module-c1-capacity-development/c1-case-studies/case-study-c15-zero-tillage-potato-integrated-pest-management-ipm-for-climate-mitigation-in-viet-nam/es/

FAO (Food and Agriculture Organization) (2018) 'Assessing the contribution of bioeconomy to countries' economies'. Available at: www.fao.org/3/I9580EN/i9580en.pdf

German Bioeconomy Council (2015) 'Bioeconomy Policy (Part II): Synopsis of national strategies'. Available at: http://biooekonomierat.de/fileadmin/Publikationen/berichte/Bioeconomy-Policy_Part-II.pdf

German Bioeconomy Council (2017) 'Bioeconomy Policy (Part I): Synopsis and analysis of strategies in the G7'. Available at: https://biooekonomierat.de/fileadmin/international/Bioeconomy-Policy_Part-I.pdf

Hasenheit, M., Gerdes, H., Kiresiewa, Z. and Beekman, V. (2016) 'Summary report on the social, economic and environmental impacts of the bioeconomy'. Available at: www.bio-step.eu/fileadmin/BioSTEP/Bio_documents/BioSTEP_D2.2_Impacts_of_the_bioeconomy.pdf

Heimann, T. (2018) 'Bioeconomy and SDGs: Does the bioeconomy support the achievement of the SDGs?', *Earth's Future*, vol. 7, no. 1, pp. 43–57.

IPBES (2019) *Global Assessment Report on Biodiversity and Ecosystem Services of the Intergovernmental Science Policy Platform on Biodiversity and Ecosystem Services*. Bonn, Germany: IPBES Secretariat.

Issa, I. (2019) 'Bioeconomy from experts' perspectives: Results of a global expert survey', *PLoS One*, vol. 14, no. 5: e0215917, doi: 10.1371/journal.pone.0215917', www.ncbi.nlm.nih.gov/pmc/articles/PMC6494193.

Juerges, N., and Hansjürgens, B. (2016) 'Soil governance in the transition towards a sustainable bioeconomy: A review'. Available at: https://doi.org/10.1016/j.jclepro.2016.10.143

Kleinschmit, D., Arts, B., Giurca, A., Mustalahti, I., Sergent, A. and Pülzl, H. (2017) 'Environmental concerns in political bioeconomy discourses', *International Forestry Review*, vol. 19, pp. 41–55.

Meyer, R. (2017) 'Bioeconomy strategies: Contexts, visions, guiding implementation principles and resulting debates', *Sustainability*, vol. 9, p. 1031

Ministry of Internal Affairs and Communications (2019) 'Community revival cooperation corps'. Available at: www.soumu.go.jp/main_content/000610488.pdf

Nagothu, U.S. (2015) 'The future of food security: Summary and recommendations', in U.S. Nagothu (ed.) *Food Security and Development: Country Case Studies* London: Earthscan.

Nahya, A. (2019) 'Finnish forest-based companies in transition to the circular bioeconomy: Drivers, organizational resources and innovations'. Available at: www.sciencedirect.com/science/article/pii/S1389934118303563?dgcid=rss_sd_all

Olivares, F.L., Busato, J.G., de Paula, A.M., Lima, L.S., Aguiar, N.O. and Canellas, L.P. (2017) 'Plant growth promoting bacteria and humic substances: Crop promotion and mechanisms of action'. Available at: https://link.springer.com/article/10.1186/s40538-017-0112-x

Pfau, S., Hagens, J., Dankbaar, B. and Smits, A. (2014) 'Visions of sustainability in bioeconomy research', *Sustainability*, vol. 6, pp. 1222–1249.

TransNational Institute (2015) 'The bioeconomy'. Available at: www.tni.org/files/publication-downloads/tni_primer_the_bioeconomy.pdf

United Nations (2019) 'UN Report: Nature's dangerous decline "unprecedented"; species extinction rates "accelerating" '. Available at: www.un.org/sustainabledevelopment/blog/2019/05/nature-decline-unprecedented-report/

WEF (World Economic Forum) (2019a) 'How extreme weather conditions could last longer due to climate change'. Available at: www.weforum.org/agenda/2019/08/extreme-weather-conditions-could-last-longer-due-climate-change/

WEF (World Economic Forum) (2019b) 'The future availability of natural resources: A new paradigm for global resource availability'. Available at: www3.weforum.org/docs/WEF_FutureAvailabilityNaturalResources_Report_2014.pdf

Wesseler, J. and von Braun, J. (2017) 'Measuring the bioeconomy: Economics and policies', *Annual Review Resource Economics*, vol. 9, pp. 275–298.

WMO (World Meteorological Organization) (2019) 'The year 2019 starts with extreme, high-impact weather'. Available at: https://public.wmo.int/en/media/news/2019-starts-extreme-high-impact-weather

WRI (World Resources Institute) (2017) 'In business, what is transformative change?' Available at: www.wri.org/blog/2017/05/business-what-transformative-change

Zhang, C., Liu, G., Xue, S. and Sun, C. (2013) 'Soil organic carbon and total nitrogen storage as affected by land use in a small watershed of the Loess Plateau, China', *European Journal of Soil Biology*, vol. 54, pp. 16–24.

Zilberman, D., Kima,E., Kirschnera, S. and Reeves, J. (2013) 'Technology and the future bioeconomy', *Agricultural Economics*, vol. 44, pp. 95–102.

Index

Page numbers in bold denote tables, those in italics denote figures.

For Product Safety Concerns and Information please contact our EU representative GPSR@taylorandfrancis.com
Taylor & Francis Verlag GmbH, Kaufingerstraße 24, 80331 München, Germany

www.ingramcontent.com/pod-product-compliance
Lightning Source LLC
LaVergne TN
LVHW011118300926
849160LV00008B/433